Modulation Theory

RIVER PUBLISHERS SERIES IN COMMUNICATIONS

Series Editors:

ABBAS JAMALIPOUR
The University of Sydney
Australia

MARINA RUGGIERI
University of Rome Tor Vergata
Italy

JUNSHAN ZHANG
Arizona State University
USA

Indexing: All books published in this series are submitted to the Web of Science Book Citation Index (BkCI), to CrossRef and to Google Scholar.

The "River Publishers Series in Communications" is a series of comprehensive academic and professional books which focus on communication and network systems. Topics range from the theory and use of systems involving all terminals, computers, and information processors to wired and wireless networks and network layouts, protocols, architectures, and implementations. Also covered are developments stemming from new market demands in systems, products, and technologies such as personal communications services, multimedia systems, enterprise networks, and optical communications.

The series includes research monographs, edited volumes, handbooks and textbooks, providing professionals, researchers, educators, and advanced students in the field with an invaluable insight into the latest research and developments.

For a list of other books in this series, visit www.riverpublishers.com

Modulation Theory

Marcelo Sampaio de Alencar

Federal University of Bahia (UFBA)
Brazil

Routledge
Taylor & Francis Group

LONDON AND NEW YORK

Published 2018 by River Publishers
River Publishers
Alsbjergvej 10, 9260 Gistrup, Denmark
www.riverpublishers.com

Distributed exclusively by Routledge
4 Park Square, Milton Park, Abingdon, Oxon OX14 4RN
605 Third Avenue, New York, NY 10017, USA

First issued in paperback 2023

Modulation Theory / by Marcelo Sampaio de Alencar.

Routledge is an imprint of the Taylor & Francis Group, an informa business

Publisher's Note
The publisher has gone to great lengths to ensure the quality of this reprint but points out that some imperfections in the original copies may be apparent.

While every effort is made to provide dependable information, the publisher, authors, and editors cannot be held responsible for any errors or omissions.

ISBN 13: 978-87-7022-942-5 (pbk)
ISBN 13: 978-87-7022-026-2 (hbk)
ISBN 13: 978-1-003-33886-4 (ebk)

This book is dedicated to my grandchildren
Vicente and Cora.

Contents

Preface

This book is intended to serve as a complementary textbook for courses dealing with Modulation Theory or Communication Systems, but also as a professional book, for engineers who need to revamp their knowledge in the area.

The modulation aspects presented in the book use modern concepts of stochastic processes, such as autocorrelation and power spectrum density, which are novel for undergraduate texts or professional books, and provide a general approach for the theory, with real life results.

The book is suitable for the undergraduate as well as the initial graduate levels of Electrical Engineering courses, and is useful for the professional who wants to review or get acquainted with the a modern exposition of the modulation theory.

Description of the Book

The books covers signal representations for most known waveforms, Fourier analysis, and presents an introduction to Fourier transform and signal spectrum, including the concepts of convolution, autocorrelation, and power spectral density, for deterministic signals, in Chapter 1.

Chapter 2 introduces the concepts of probability, random variables, and stochastic processes, including autocorrelation, cross-correlation, power spectral and cross-spectral densities, for random signals, and their applications to the analysis of linear systems. This chapter also includes the response of specific non-linear systems, such as power amplifiers.

The book presents amplitude modulation with random signals, including analog and digital signals, and discusses performance evaluation methods, in Chapter 3.

The subject of Chapter 4 is quadrature amplitude modulation using random signals. Several modulation schemes are discussed, including SSB, QAM, ISB, C-QUAM, QPSK, and MSK. Their autocorrelation and power spectrum densities are computed.

Chapter 5 discusses angle modulation with random modulating signals, along with frequency and phase modulation, and orthogonal frequency division multiplexing. Their power spectrum densities are computed using the Wiener–Khintchin theorem.

The characteristics of digital modulation techniques are discussed in Chapter 6. The book has two appendices, an appendix that summarizes the main formulas, and another one that presents and explains the usual communication acronyms.

List of Figures

1

Theory of Signals and Linear Systems

1.1 Introduction

This chapter provides the necessary mathematical basis for the reader to understand the modulation theory (Alencar, 1999). There is a basic introduction to signal analysis, in which the most commonly used signals are presented (Baskakov, 1986). The main concepts associated with Fourier series and Fourier transform are introduced. The theory and the properties of the Fourier transform are presented, along with the main properties of signal and system analysis (Papoulis, 1983b).

1.2 Signal Analysis

1.2.1 Linearity

A communication system is a complex combination of filters, modulators, demodulators, attenuators, amplifiers, samplers, quantizers, coders, equalizers, oscillators, and other analog and digital equipment, to process, transmit, and receive signals.

It is a usual practice to break the system into parts, or analyze it, to obtain the necessary design information. Most devices are non-linear by nature, which makes it difficult to model them, but they can be linearized to a certain extent, if the operational range of the component is restricted, for example, in frequency, time, or power.

Linearity is an important property when studying communication systems. A system is defined to be linear if it satisfies the properties of homogeneity and additivity.

1. Homogeneity – If the application of the signal $x(t)$ at the system input produces $y(t)$ at the system output, then the application of the input $\alpha x(t)$, in which α is a constant, produces $\alpha y(t)$ at the output.
2. Additivity – If the individual application of the signals $x_1(t)$ and $x_2(t)$ at the system input produces, respectively, $y_1(t)$ and $y_2(t)$ at the system

1

output, then the joint application of the input $x_1(t) + x_2(t)$ produces $y_1(t) + y_2(t)$ at the output.

1.2.2 The Convolution Theorem

The convolution between two time functions $f(t)$ and $g(t)$ is defined by the following integral

$$h(t) = \int_{-\infty}^{\infty} f(\tau)g(t - \tau)d\tau, \tag{1.1}$$

which is often denoted as $h(t) = f(t) * g(t)$.

The convolution is a linear operation in the time domain, and can be used to obtain the response of a linear system, given the input signal and the impulse response of the system.

1.3 Some Important Functions

This section presents the characteristics of some functions that are commonly used in signal analysis.

1.3.1 The Constant Function

The constant function is invariable in time; it can be written as $f(t) = A$, for $-\infty < t < \infty$. It is an idealization of a real constant signal that needs to be switched on some time, but it is useful if one considers that the signal source has been in operation a sufficient time for the signal to stabilize completely.

1.3.2 The Sine and the Cosine Functions

The sinusoidal functions, sine and cosine, can represent an analog signal, in an electronic circuit, a signal that propagates though space, a communication carrier, the series of harmonics of a complex signal, the output of an oscillator, and so on.

$$x(t) = A \cos(\omega t + \phi) \tag{1.2}$$

The sinusoidal function, illustrated in Figure 1.1, is characterized by its initial phase ϕ, its spectral frequency ω, and its amplitude A. The period of the sinusoid, the interval between two maxima of the waveform, is given by $T = 1/f = 2\pi/\omega$, in which f is the frequency, in hertz.

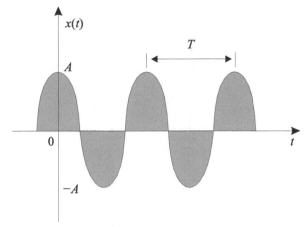

Figure 1.1 The sine function.

1.3.3 The Heaviside Step Function

The unit step function $u(t)$, also called Heaviside step, is considered a generalized function, because it does not comply with the requisites to be a regular continuous function. It was invented by Oliver Heaviside (1850–1925), an English scientist, to model the switching of voltage or current sources, for example, but can be used in several other instances.

The unit step function is defined as

$$u(t) = \begin{cases} 1 & \text{if } t \geq 0 \\ 0 & \text{if } t < 0 \end{cases} \tag{1.3}$$

and illustrated in Figure 1.2.

Example: Compute the convolution of functions, $h(t) = f(t) * g(t)$, for $f(t) = g(t) = u(t)$.

Solution: Using the definition of convolution, from Formula (1.1), one obtains

$$h(t) = \int_{-\infty}^{\infty} u(\tau)u(t - \tau)d\tau, \tag{1.4}$$

which, because of the limiting property of the unit step function, can be written as

$$h(t) = \int_{0}^{t} u(\tau)u(t - \tau)d\tau, \tag{1.5}$$

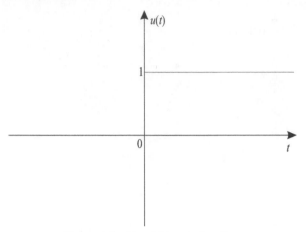

Figure 1.2   Heaviside step function.

or

$$h(t) = \int_0^t d\tau = t, \text{ for } t \geq 0, \text{ and zero otherwise.} \quad (1.6)$$

This example can be used as the definition of the ramp function $r(t)$.

The unit step function can be used to create other functions, as composite functions, in a simple and compact manner. Consider a generalization of Formula (1.3), using function composition,

$$u(f(t)) = \begin{cases} 1 & \text{if } f(t) \geq 0 \\ 0 & \text{if } f(t) < 0 \end{cases} \quad (1.7)$$

then the domain of the step function is the image of $f(t)$, and this gives a poetic license to create several new signals. For example, if $f(t) = \cos(\omega t + \phi)$, the application of Formula (1.7) gives

$$u(\cos(\omega t + \phi)) = \begin{cases} 1 & \text{if } \cos(\omega t + \phi) \geq 0 \\ 0 & \text{if } \cos(\omega t + \phi) < 0 \end{cases} \quad (1.8)$$

which models a binary periodic signal, continuous in time, with amplitude levels in the set $\{0, 1\}$.

1.3.4 The Ramp Function

The ramp function, illustrated in Figure 1.3, is the integral, or primitive, of the Heaviside step function. It is null, from minus infinity to zero, and it is

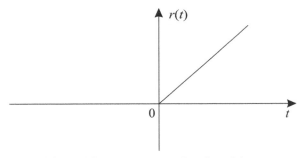

Figure 1.3   The ramp as a function of time.

equal to the identity function, from zero to plus infinity.

$$r(t) = \int_{-\infty}^{t} u(t)dt. \tag{1.9}$$

The ramp function can be defined as

$$r(t) = \begin{cases} t & \text{if } t \geq 0 \\ 0 & \text{if } t < 0 \end{cases} \tag{1.10}$$

The ramp function models, for example, the increasing voltage in a capacitor that is driven by a current source. The modulus function can be written with the aid of the ramp function, as $|t| = r(t) + r(-t)$.

The ramp function can also be very useful to define other functions in a simple manner. Suppose the ramp is defined as a function of time, $x(t)$, instead of time directly, as in the following,

$$r(x(t)) = \begin{cases} x(t) & \text{if } x(t) \geq 0 \\ 0 & \text{if } t < 0 \end{cases} \tag{1.11}$$

then a new set of functions can be derived.

As an example, if the internal function is $x(t) = \sin(\omega t + \phi)$, the resulting composite function gives

$$g_R(t) = r(A\sin(\omega t + \phi)) = \begin{cases} A\sin(\omega t + \phi) & \text{if } A\sin(\omega t + \phi) \geq 0 \\ 0 & \text{if } t < 0 \end{cases} \tag{1.12}$$

which represents a rectified sine wave, as illustrated in Figure 1.4.

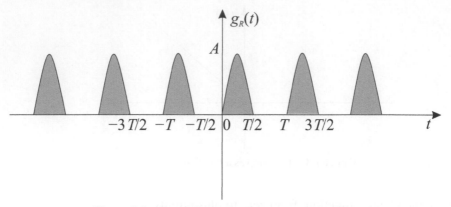

Figure 1.4   Signal obtained at the output of a rectifier.

1.3.5 The Gate Function

The gate, or pulse, function offer a somehow more realistic model for a signal that has to be switched on and off. The gate function, shown in Figure 1.5, is defined by the expression

$$p_T(t) = [u(t + T/2) - u(t - T/2)], \qquad (1.13)$$

or

$$p_T(t) = \begin{cases} 1 & \text{if } |t| \leq T/2 \\ 0 & \text{if } |t| > T/2 \end{cases} \qquad (1.14)$$

This function is useful to define operational intervals, for example, if a sinusoidal signal has to be switched on during the interval T, just multiply the signal by the gate function.

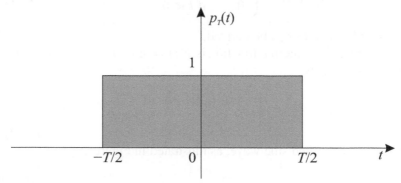

Figure 1.5   The gate or pulse function.

The gate function can be used to limit the interval of a periodic signal. For example, $g_T(t) = A\cos(\omega t + \phi) \cdot p_T(t)$ represents the windowing of the cosine function in the interval $[-T/2, T/2]$, as illustrated in Figure 1.6.

1.3.6 Impulse Function or Dirac's Delta Function

The impulse is also a generalized function that can be obtained from the unit step function by derivation, that is, $\delta = u'(t)$. In other words,

$$\delta(t) = \begin{cases} \infty & \text{if } t = 0 \\ 0 & \text{if } t \neq 0 \end{cases} \tag{1.15}$$

The impulse function, shown in Figure 1.7, has unit area,

$$\int_{-\infty}^{\infty} \delta(t) dt = 1, \tag{1.16}$$

and is useful to obtain the response of linear systems, to represent phenomena such as atmospheric lightning, capacitive or inductive discharges, to aid in the sampling operation, and many other applications.

Because the impulse is a rather strange function, the formal way to express it is inside an integral. It can be written as a kernel of the following integral, which also serves as a formal definition of the impulse.

$$\int_{-\infty}^{\infty} \delta(t) f(t) dt = f(0). \tag{1.17}$$

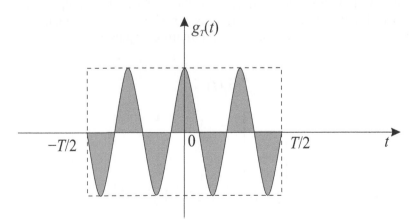

Figure 1.6 Windowing of the cosine function.

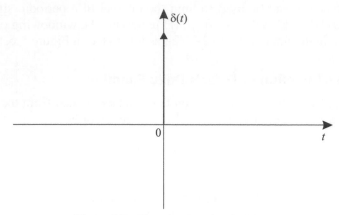

Figure 1.7   The Dirac impulse function.

This representation provides a way to present the sampling property of the impulse function,

$$\int_{-\infty}^{\infty} \delta(t - \tau)f(t)dt = f(\tau).\qquad(1.18)$$

Example: A periodic series of impulses, which is used to take repetitive samples of a signal, can be compactly written as

$$\delta_T(t) = \delta(\sin(2\pi ft)),$$

in which the impulses occur at the points $t = k/2f$, as shown in Figure 1.8.

Because the impulse is the derivative of the unit step function, it is not difficult to verify that, if $x(t) = u(\sin t)$ is a periodic binary signal, then, $y(t) = x'(t) = \cos \cdot \delta(\sin t)$ is a series of alternating impulses.

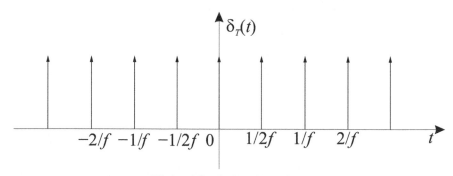

Figure 1.8   Series of impulses.

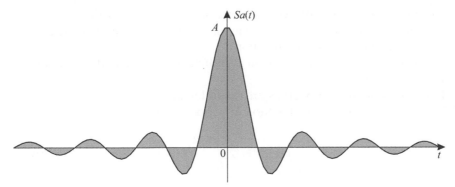

Figure 1.9   The sampling function.

1.3.7 The Sampling Function

The sampling function

$$\text{Sa}\,(\omega t) = A\frac{\sin \omega t}{\omega t} \tag{1.19}$$

appears in several instances in communications and signal analysis. This function converges to A, as t goes to zero, and goes to zero, as t approaches infinity.

The first zero of the sampling function, as shown in Figure 1.9, occurs at $\omega = 2\pi/T$, if T is the period of the sine function. The parameter ω controls the compacting and dilatation of the function along the time axis.

1.3.8 Even and Odd Functions

Mathematical functions can be either odd or even. A function is odd if it is anti-symmetric with respect to the ordinate axis, that is, if $f(-t) = -f(t)$, in which $-t$ and t are assumed to belong to the function domain. Examples of odd functions are provided by the functions t, t^3, $\sin t$, and $t^{|2n+1|}$.

Similarly, an even function is symmetric with respect to the ordinate axis, that is, if $f(-t) = f(t)$, in which t and $-t$ are assumed to belong to the function domain. Examples of even functions are provided by the functions 1, t^2, $\cos t$, $|t|$, $\exp(-|t|)$, and $t^{|2n|}$.

1.3.9 Some Elementary Properties of Functions

1. The sum (difference) of two even functions is an even function.
2. The product (quotient) of two even functions is an even function.
3. The sum (difference) of two odd functions is an odd function.

4. The product (quotient) of two odd functions is an even function.
5. The sum (difference) of an even function and an odd function is neither an even function nor an odd function.
6. The product (quotient) between an even function and an odd function is an odd function.

   Two other important properties are the following.
7. If $f(t)$ is an even periodic function of period T, then

$$\int_{-T/2}^{T/2} f(t)dt = 2 \int_{0}^{T/2} f(t)dt. \tag{1.20}$$

8. If $f(t)$ is an odd periodic function of period T, then

$$\int_{-T/2}^{T/2} f(t)dt = 0. \tag{1.21}$$

1.4 Basic Fourier Analysis

The Fourier theory establishes fundamental conditions for the representation of an arbitrary function in a finite interval as a sum of sinuoids. In fact, this is just an instance of the more general Fourier representation of signals in which a periodic signal $f(t)$, under fairly general conditions, can be represented by a complete set of orthogonal functions.

By a complete set $\mathcal{F}$ of orthogonal functions, it is understood that, except for those orthogonal functions already in $\mathcal{F}$, there are no other orthogonal functions that belong to $\mathcal{F}$ to be considered.

The periodic signal $f(t)$ must satisfy the Dirichlet conditions, that is, $f(t)$ is a bounded function which in any one period has at most a finite number of local maxima and minima and a finite number of points of discontinuity (Wylie, 1966).

The representation of signals by orthogonal functions usually presents an error, which diminishes as the number of component terms in the corresponding series is increased. This error produces the Gibbs phenomenon, an oscillation that occurs at the transition points (Schwartz and Shaw, 1975).

The expansion of a periodic signal $f(t)$ as a sum of mutually orthogonal functions requires a review of the concepts of periodicity and orthogonality.

A given function $f(t)$ is periodic, of period T, as illustrated in Figure 1.10, if and only if, T is the smallest positive number for which $f(t + T) = f(t)$. In other words, $f(t)$ is periodic if its domain contains $t + T$ whenever it contains t, and $f(t + T) = f(t)$.

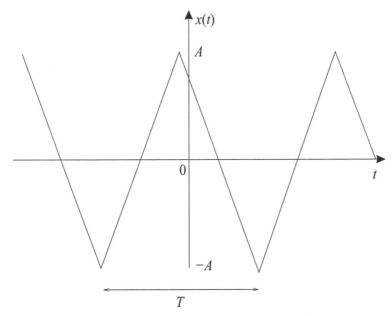

Figure 1.10   Example of a periodic signal.

It follows from the definition of a periodic function that if T represents the period of $f(t)$, then $f(t) = f(t+nT)$, for $n = 1, 2, \ldots$, ,that is, $f(t)$ repeats its values when integer multiples of T (Wozencraft and Jacobs, 1965a) are added to its argument.

If $f(t)$ and $g(t)$ are two periodic functions with the same period T, then their sum $f(t)+g(t)$ will also be a periodic function with period T. This result can be proven if one makes $h(t) = f(t) + g(t)$, and notices that $h(t + T) - f(t + T) + g(t + T) = f(t) + g(t) = h(t)$.

Orthogonality provides the tool to introduce the concept of a basis, that is, a minimum set of functions that can be used to generate other functions. However, orthogonality by itself does not guarantee that a complete vector space is generated.

Two real functions $u(t)$ and $v(t)$, defined in the interval $\alpha \le t \le \beta$, are orthogonal if their inner product is null, that is, if

$$< u(t), v(t) > = \int_{\alpha}^{\beta} u(t)v(t)dt = 0. \tag{1.22}$$

The set of Walsh functions $w_n(t)$, as illustrated in Figure 1.11, can be used to represent signals in the time domain. This set of functions constitutes an orthogonal set in the interval $(0, 1)$.

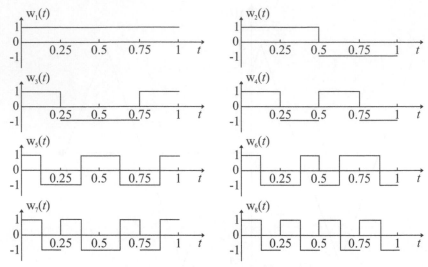

Figure 1.11   Set of orthogonal functions.

1.4.1 The Trigonometric Fourier Series

The trigonometric Fourier series representation of the signal $f(t)$ can be written as

$$f(t) = a_0 + \sum_{n=1}^{\infty} [a_n \cos(n\omega_0 t) + b_n \sin(n\omega_0 t)], \qquad (1.23)$$

in which the term a_0 (the average value of the function $f(t)$) indicates whether or not the signal contains a constant value and the terms a_n and b_n are called the Fourier series coefficients, in which n is a positive integer. The equality sign holds in (1.23) for all values of t only when $f(t)$ is periodic.

The Fourier series representation is a useful tool for any type of signal, as long as that signal representation is required only in the $[0, T]$ interval. Outside that interval, the Fourier series representation is always be periodic, even if the signal $f(t)$ is not periodic (Knopp, 1990).

The trigonometric are examples of orthogonal functions because they satisfy the following equations, called orthogonality relations, for integer values of n and m:

$$\int_0^T \cos(n\omega_0 t) \sin(m\omega_0 t)dt = 0, \text{ for all integers } n, m, \qquad (1.24)$$

$$\int_0^T \cos(n\omega_o t)\cos(m\omega_o t)dt = \begin{cases} 0 & \text{if } n \neq m \\ \frac{T}{2} & \text{if } n = m \end{cases} \qquad (1.25)$$

$$\int_0^T \sin(n\omega_o t)\sin(m\omega_o t)dt = \begin{cases} 0 & \text{if } n \neq m \\ \frac{T}{2} & \text{if } n = m \end{cases} \qquad (1.26)$$

in which $\omega_0 = 2\pi/T$.

As a consequence of the orthogonality conditions, explicit expressions for the coefficients a_n and b_n of the Fourier trigonometric series can be computed. By integrating both sides in expression (1.23) in the interval $[0, T]$, it follows that (Oberhettinger, 1990)

$$\int_0^T f(t)dt = \int_0^T a_o dt + \sum_{n=1}^{\infty} \int_0^T a_n \cos(n\omega_o t)dt + \sum_{n=1}^{\infty} \int_0^T b_n \sin(n\omega_o t)dt$$

and since

$$\int_0^T a_n \cos(n\omega_o t)dt = \int_0^T b_n \sin(n\omega_o t)dt = 0,$$

it follows that

$$a_o = \frac{1}{T}\int_0^T f(t)dt. \qquad (1.27)$$

Multiplication of both sides in expression (1.23) by $\cos(m\omega_o t)$, and integrating in the interval $[0, T]$, leads to

$$\int_0^T f(t)\cos(m\omega_o t)dt = \int_0^T a_o \cos(m\omega_o t)dt \qquad (1.28)$$

$$+ \sum_{n=1}^{\infty} \int_o^T a_n \cos(n\omega_o t)\cos(m\omega_o t)dt$$

$$+ \sum_{n=1}^{\infty} \int_o^T b_n \cos(m\omega_0 t)\sin(n\omega_o t)dt,$$

which, after simplification, gives

$$a_n = \frac{2}{T}\int_0^T f(t)\cos(n\omega_o t)dt, \text{ for } n = 1, 2, 3, \ldots \qquad (1.29)$$

In a similar way, b_n is found by multiplying both sides in expression (1.23) by $\sin(n\omega_o t)$ and integrating in the interval $[0, T]$, that is,

$$b_n = \frac{2}{T} \int_0^T f(t) \sin(n\omega_o t)dt, \qquad (1.30)$$

for $n = 1, 2, 3, \ldots$.

A considerable simplification, when computing coefficients of a trigonometric Fourier series, is obtained with properties (7) and (8):

1. If $f(t)$ is an even function, then $b_n = 0$, and

$$a_n = \frac{2}{T} \int_0^T f(t) \cos(n\omega_o t)dt, \quad \text{for } n = 1,2,3,\ldots \qquad (1.31)$$

2. If $f(t)$ is an odd function, then $a_n = 0$ and

$$b_n = \frac{2}{T} \int_0^T f(t) \sin(n\omega_o t)dt, \quad \text{for } n = 1,2,3,\ldots \qquad (1.32)$$

Example: Compute the coefficients of the trigonometric Fourier series for the waveform $f(t) = A[u(\cos(2\pi f t)]$, which repeats itself with period $T = 1/f$, in which $u(t)$ denotes the unit step function.

Solution: Since the given signal is symmetric with respect to the ordinate axis, it follows that $f(t) = f(-t)$ and the function is even. Therefore, $b_n = 0$, and all that is left to compute is a_o, and a_n for $n = 1, 2, \ldots$.

The expression to calculate the average value a_0 is

$$a_o = \frac{1}{T} \int_{-\frac{T}{2}}^{\frac{T}{2}} f(t)dt = \frac{1}{T} \int_{-\tau}^{\tau} Adt = \frac{2A\tau}{T}.$$

In the previous equation, the maximum value of τ is $T/2$. The coefficients a_n for $n = 1, 2, \ldots$ are computed as

$$a_n = \frac{2}{T} \int_0^T f(t) \cos(n\omega_o t)dt = \frac{2}{T} \int_{-\tau}^{\tau} A \cos(n\omega_o t)dt,$$

$$a_n = \frac{4A}{T} \int_0^\tau \cos(n\omega_o t)dt = \frac{4A}{Tn\omega_o} \sin(n\omega_o t) \Big|_0^\tau = (4A\tau/T)\frac{\sin(n\omega_o \tau)}{n\omega_0 \tau}.$$

Therefore, the signal $f(t)$ is represented by the following trigonometric Fourier series

$$f(t) = \frac{2A\tau}{T} + \left(\frac{4A\tau}{T}\right) \sum_{n=1}^{\infty} \frac{\sin(n\omega_o \tau)}{n\omega_0 \tau} \cos(n\omega_o t).$$

1.4.2 The Compact Fourier Series

It is also possible to represent the Fourier series in a form known as the compact Fourier series as follows

$$f(t) = C_0 + \sum_{n=1}^{\infty} C_n \cos(n\omega_o t + \theta_n). \tag{1.33}$$

By expanding the expression $C_n \cos(n\omega_o t + \theta)$ as $C_n \cos(n\omega_o t) \cos \theta_n - C_n \sin(n\omega_o t) \sin \theta_n$ and comparing this result with (1.23), it follows that $a_o = C_o$, $a_n = C_n \cos \theta_n$, and $b_n = -C_n \sin \theta_n$. It is now possible to compute C_n as a function of a_n and b_n. For that purpose, it is sufficient to square a_n and b_n and add the result, that is,

$$a_n^2 + b_n^2 = C_n^2 \cos^2 \theta_n + C_n^2 \sin^2 \theta_n = C_n^2. \tag{1.34}$$

From Equation (1.34), the modulus of C_n can be written as

$$C_n = \sqrt{a_n^2 + b_n^2}. \tag{1.35}$$

In order to determine θ_n, it suffices to divide b_n by a_n, that is,

$$\frac{b_n}{a_n} = -\frac{\sin \theta_n}{\cos \theta_n} = -\tan \theta_n, \tag{1.36}$$

which, when solved for θ_n, gives

$$\theta_n = -\arctan\left(\frac{b_n}{a_n}\right). \tag{1.37}$$

1.4.3 The Exponential Fourier Series

Since the set of exponential functions $e^{jn\omega_o t}$, $n = 0, \pm1, \pm2, \ldots$, is a complete set of orthogonal functions in an interval of magnitude T, in which $T = 2\pi/\omega_o$, then it is possible to represent a function $f(t)$ by a linear combination of exponential functions in an interval T.

$$f(t) = \sum_{-\infty}^{\infty} F_n e^{jn\omega_o t} \tag{1.38}$$

in which

$$F_n = \frac{1}{T} \int_{\frac{-T}{2}}^{\frac{T}{2}} f(t) e^{-jn\omega_o t} dt. \tag{1.39}$$

Equation (1.38) represents the exponential Fourier series expansion of $f(t)$ and Equation (1.39) is the expression to compute the associated series coefficients. The exponential Fourier series is also known as the complex Fourier series. It can be shown that Equation (1.38) is just another way of expressing the Fourier series as given in (1.23). Replacing $\cos(nw_ot) + j\sin(nw_ot)$ for e^{nw_ot} (Euler's identity) in (1.38), it follows that

$$f(t) = F_o + \sum_{n=-\infty}^{-1} F_n[\cos(nw_ot) + j\sin(nw_ot)]$$

$$+ \sum_{n=1}^{\infty} F_n[\cos(nw_ot) + j\sin(nw_ot)],$$

or

$$f(t) = F_o + \sum_{n=1}^{\infty} F_n[\cos(nw_ot)+j\sin(nw_ot)]+F_{-n}[\cos(nw_ot)-j\sin(nw_ot)].$$

After grouping the coefficients of the sine and cosine terms, it follows that

$$f(t) = F_o + \sum_{n=1}^{\infty}(F_n + F_{-n})\cos(nw_ot) + j(F_n - F_{-n})\sin(nw_ot). \quad (1.40)$$

Comparing the this expression with Expression (1.23) it follows that

$$a_o = F_o, \ a_n = (F_n + F_{-n}) \text{ and } b_n = j(F_n - F_{-n}), \quad (1.41)$$

and that

$$F_o = a_o, \quad (1.42)$$

$$F_n = \frac{a_n - jb_n}{2}, \quad (1.43)$$

and

$$F_{-n} = \frac{a_n + jb_n}{2}. \quad (1.44)$$

In case the function $f(t)$ is even, that is, if $b_n = 0$, then

$$a_o = F_o, \ F_n = \frac{a_n}{2}, \text{ and } F_{-n} = \frac{a_n}{2}. \quad (1.45)$$

Example: Compute the exponential Fourier series for the train of impulses given by,

$$\delta_T(t) = \delta[\sin(2\pi f t)], \ f = 1/T.$$

Solution: The complex coefficients are given by

$$F_n = \frac{1}{T} \int_{\frac{-T}{2}}^{\frac{T}{2}} \delta_T(t) e^{-jn\omega_o t} dt = \frac{1}{T}, \tag{1.46}$$

since

$$\int_{-\infty}^{\infty} \delta(t - t_o) f(t) dt = f(t_o), \text{ (property of impulse filtering).} \tag{1.47}$$

Observe that $f(t)$ can be written as

$$f(t) = \frac{1}{T} \sum_{n=-\infty}^{\infty} e^{-jn\omega_o t}. \tag{1.48}$$

The impulse train is an idealization, as most functions are, of a real signal. In practice, in order to obtain an impulse train, it is sufficient to pass a binary digital signal through a differentiator circuit and then pass the resulting waveform through a half-wave rectifier.

The Fourier series expansion of a periodic signal is equivalent to its decomposition in frequency components. In general, a periodic function with period T has frequency components $0, \pm\omega_o, \pm 2\omega_o, \pm 3\omega_o, \ldots, \pm n\omega_o$, in which $\omega_o = 2\pi/T$ is the fundamental frequency and the multiples of ω_0 are called harmonics. Notice that the spectrum exists only for discrete values of ω and that the spectral components are spaced by at least ω_o.

1.5 Fourier Transform

It has been shown that an arbitrary function can be represented in terms of an exponential (or trigonometric) Fourier series in a finite interval. If such a function is periodic, this representation can be extended for the entire interval $(-\infty, \infty)$.

However, it is interesting to observe the spectral behavior of a function in general, periodic or not, in the entire interval $(-\infty, \infty)$. To do that, it is necessary to truncate the function $f(t)$ in the interval $[-T/2, T/2]$, to obtain $f_T(t)$. It is possible then to represent this function as a sum of exponentials in the entire interval $(-\infty, \infty)$ if T goes to infinity, as follows.

$$\lim_{T \to \infty} f_T(t) = f(t).$$

The $f_T(t)$ signal can be represented by the exponential Fourier series as

$$f_T(t) = \sum_{n=-\infty}^{\infty} F_n e^{jn\omega_o t}, \tag{1.49}$$

in which $\omega_o = 2\pi/T$ and

$$F_n = \frac{1}{T} \int_{-\frac{T}{2}}^{\frac{T}{2}} f_T(t) e^{-jn\omega_o t} dt. \tag{1.50}$$

The coefficients F_n represent the spectral amplitude associated to each component of frequency $n\omega_o$. As T increases, the amplitudes diminish but the spectrum shape is not altered. The increase in T forces ω_o to diminish and the spectrum to become denser. In the limit, as $T \to \infty$, ω_o becomes infinitesimally small, being represented by $d\omega$. On the other hand, there are now infinitely many components and the spectrum is no longer a discrete one, becoming a continuous spectrum in the limit.

For convenience, write $TF_n = F(\omega)$, that is, the product TF_n becomes a function of the variable ω, since $n\omega_o \to \omega$. Replacing $\frac{F(\omega)}{T}$ for F_n in (1.49), one obtains

$$f_T(t) = \frac{1}{T} \sum_{n=-\infty}^{\infty} F(\omega) e^{j\omega t}. \tag{1.51}$$

Replacing $\omega_0/2\pi$ for $1/T$,

$$f_T(t) = \frac{1}{2\pi} \sum_{n=-\infty}^{\infty} F(\omega) e^{j\omega t} \omega_0. \tag{1.52}$$

In the limit, as T approaches infinity, one has

$$f(t) = \frac{1}{2\pi} \int_{-\infty}^{\infty} F(\omega) e^{j\omega t} d\omega \tag{1.53}$$

which is known as the inverse Fourier transform.

Similarly, from (1.50), as T approaches infinity, one obtains

$$F(\omega) = \int_{-\infty}^{\infty} f(t) e^{-j\omega t} dt \tag{1.54}$$

which is known as the direct Fourier transform, sometimes denoted in the literature as $F(\omega) = \mathcal{F}[f(t)]$. A Fourier transform pair is often denoted as $f(t) \longleftrightarrow F(\omega)$.

Some important Fourier transforms are presented in the following (Haykin, 1988).

1.5.1 Bilateral Exponential Signal

If $f(t) = e^{-a|t|}$, it follows from Formula (1.54) that

$$F(\omega) = \int_{-\infty}^{\infty} e^{-a|t|} e^{-j\omega t} dt \tag{1.55}$$

$$= \int_{-\infty}^{0} e^{at} e^{-j\omega t} dt + \int_{0}^{\infty} e^{-at} e^{-j\omega t} dt$$

$$= \frac{1}{a - j\omega} + \frac{1}{a + j\omega},$$

$$F(\omega) = \frac{2a}{a^2 + \omega^2}. \tag{1.56}$$

1.5.2 Transform of the Gate Function

The Fourier transform of the gate function, illustrated in Figure 1.5, can be calculated as follows.

$$F(\omega) = \int_{-\frac{T}{2}}^{\frac{T}{2}} A e^{-j\omega t} dt \tag{1.57}$$

$$= \frac{A}{j\omega} (e^{j\omega \frac{T}{2}} - e^{-j\omega \frac{T}{2}})$$

$$= \frac{A}{j\omega} 2j\sin(\omega T/2),$$

which can be rearranged as

$$F(\omega) = AT \left(\frac{\sin(\omega T/2)}{\omega T/2} \right),$$

and finally

$$F(\omega) = AT \text{Sa} \left(\frac{\omega T}{2} \right), \tag{1.58}$$

in which $\mathrm{Sa}\,(x) = \frac{\sin x}{x}$ is the sampling function. This function converges to one, as x goes to zero. The sampling function, the magnitude of which is illustrated in Figure 1.12, is of great relevance in communication theory.

The sampling function obeys the following important relationship

$$\int_{-\infty}^{\infty} \frac{k}{\pi} \mathrm{Sa}\,(kt)dt = 1. \qquad (1.59)$$

The area under this curve is equal to 1. As k increases, the amplitude of the sampling function increases, the spacing between zero crossings diminishes, and most of the signal energy concentrates near the origin. For $k \to \infty$, the function converges to an impulse function, that is.

$$\delta(t) = \lim_{k \to \infty} \frac{k}{\pi} \mathrm{Sa}\,(kt). \qquad (1.60)$$

In this manner, in the limit, it is true that $\int_{-\infty}^{\infty} \delta(t)dt = 1$. Since the function concentrates its non-zero values near the origin, it follows that $\delta(t) = 0$ for $t \neq 0$.

Equation (1.60), which is a limiting property of the sampling function, can be used as a definition of the impulse. In fact, several functions that have unit area, such as the Gaussian function, can also be used to define the impulse.

1.5.3 Fourier Transform of the Impulse Function

In Formula (1.54), by making the substitution $f(t) = \delta(t)$, it follows that

$$F(\omega) = \int_{-\infty}^{\infty} \delta(t)e^{-j\omega t}dt. \qquad (1.61)$$

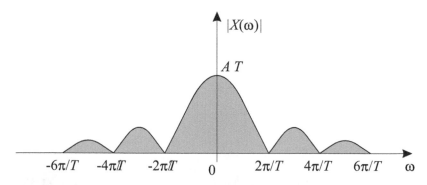

Figure 1.12   Magnitude plot of the Fourier transform of the gate function.

Using the impulse filtering property, one obtains $F(\omega) = 1$. Because the Fourier transform is constant, over the entire spectrum, one concludes that the impulse function contains a continuum of equal amplitude spectral components.

An impulse can model a sudden electrostatic discharge, such as the lightning that occurs during a thunderstorm. The effect of the lightning can be felt, for example, in several radiofrequency ranges, including the AM and FM radio bands, and the TV band.

Alternatively, by making $F(\omega) = 1$ in (1.53) and simplifying, the impulse function can be written as

$$\delta(t) = \frac{1}{\pi} \int_0^\infty \cos \omega t d\omega.$$

1.5.4 Transform of the Constant Function

It is necessary a word of caution in this case. If $f(t)$ is a constant function, then its Fourier transform in principle would not exist since this function does not satisfy the absolute integrability criterion. In general, $F(\omega)$, the Fourier transform of $f(t)$, is expected to be finite, that is,

$$|F(\omega)| \leq \int_{-\infty}^\infty |f(t)||e^{-j\omega t}|dt < \infty, \tag{1.62}$$

since $|e^{-j\omega t}| = 1$, then

$$\int_{-\infty}^\infty |f(t)|dt < \infty. \tag{1.63}$$

However, that is just a sufficiency condition and not a necessary condition for the existence of the Fourier transform, since there exist functions that although do not satisfy the condition of absolute integrability, in the limit have a Fourier transform (Carlson, 1975).

This observation is important, since this approach is often used to compute Fourier transforms of several functions. The constant function can be approximated by a gate function with amplitude A and width τ, if τ approaches the infinity,

$$\mathcal{F}[A] = \lim_{\tau \to \infty} A\tau \mathrm{Sa}\left(\frac{\omega\tau}{2}\right) \tag{1.64}$$

$$= 2\pi A \lim_{\tau \to \infty} \frac{\tau}{2\pi} \mathrm{Sa}\left(\frac{\omega\tau}{2}\right)$$

$$\mathcal{F}[A] = 2\pi A\delta(\omega). \tag{1.65}$$

This result is interesting and also intuitive, since a constant function in time represents a DC level and, as was to be expected, contains no spectral component except for the one at $\omega = 0$.

1.5.5 Fourier Transform of the Sine and Cosine Functions

Since both the sine and the cosine functions are periodic functions, they do not satisfy the condition of absolute integrability. However, their respective Fourier transforms exist in the limit when τ goes to infinity.

Assuming the function to exist only in the interval $(\frac{-\tau}{2}, \frac{\tau}{2})$ and to be zero outside this interval, and considering the limit of the expression when τ goes to infinity,

$$\mathcal{F}(\sin \omega_0 t) = \lim_{\tau \to \infty} \int_{\frac{-\tau}{2}}^{\frac{\tau}{2}} \sin \omega_0 t \; e^{-j\omega t} dt \tag{1.66}$$

$$= \lim_{\tau \to \infty} \int_{\frac{-\tau}{2}}^{\frac{\tau}{2}} \frac{e^{-j(\omega-\omega_0)t}}{2j} - \frac{e^{-j(\omega+\omega_0)t}}{2j} dt$$

$$= \lim_{\tau \to \infty} \left[\frac{j\tau \sin(\omega + \omega_0)\frac{\tau}{2}}{2(\omega + \omega_0)\frac{\tau}{2}} - \frac{j\tau \sin(\omega - \omega_0)\frac{\tau}{2}}{2(\omega - \omega_0)\frac{\tau}{2}} \right]$$

$$= \lim_{\tau \to \infty} \left\{ j\frac{\tau}{2}\mathrm{Sa}\left[\frac{(\omega + \omega_0)}{2}\right] - j\frac{\tau}{2}\mathrm{Sa}\left[\frac{\tau(\omega + \omega_0)}{2}\right] \right\}.$$

Therefore,

$$\mathcal{F}(\sin\omega_0 t) = j\pi[\delta(\omega + \omega_0) - \delta(\omega - \omega_0)].$$

Applying a similar reasoning, it follows that

$$\mathcal{F}(\cos \omega_0 t) = \pi[\delta(\omega - \omega_0) + \delta(\omega + \omega_0)], \tag{1.67}$$

which is shown in Figure 1.13.

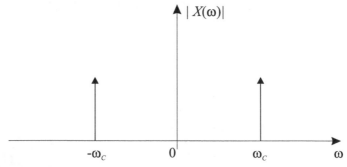

Figure 1.13   Magnitude plot for the Fourier transform of the sine function.

1.5.6 Fourier Transform of the Complex Exponential

The Fourier transform can be obtained using Euler's identity, $e^{j\omega_0 t} = \cos \omega_0 t + j\sin\omega_0 t$, and a property, as follows

$$\mathcal{F}[e^{j\omega_0 t}] = \mathcal{F}[\cos \omega_0 t + j\sin\omega_0 t]. \tag{1.68}$$

Substituting in (1.68) the Fourier transforms of the sine and of the cosine functions, respectively, it follows that

$$\mathcal{F}[e^{j\omega_0 t}] = 2\pi\delta(\omega - \omega_0). \tag{1.69}$$

1.5.7 Fourier Transform of a Periodic Function

Consider the exponential Fourier series representation of a periodic function $f_T(t)$ of period T,

$$f_T(t) = \sum_{n=-\infty}^{\infty} F_n e^{jn\omega_0 t}. \tag{1.70}$$

Applying the Fourier transform Fourier to both sides in Equation (1.70), it follows that

$$\mathcal{F}[f_T(t)] = \mathcal{F}\left[\sum_{n=-\infty}^{\infty} F_n e^{jn\omega_0 t}\right] \tag{1.71}$$

$$= \sum_{n=-\infty}^{\infty} F_n \mathcal{F}[e^{jn\omega_0 t}]. \tag{1.72}$$

Applying the result from (1.69) in (1.72), it follows that

$$F(\omega) = 2\pi \sum_{n=-\infty}^{\infty} F_n \delta(\omega - n\omega_0). \tag{1.73}$$

1.6 Some Properties of the Fourier Transform

It is possible to obtain a set of properties to help solve problems that require the use of the Fourier transform.

1.6.1 Linearity of the Fourier Transform

The application of the criteria for homogeneity and additivity makes it possible to check that the process that generates the signal $s(t) = A \cos(\omega_c t +$

$\Delta m(t) + \theta)$, from an input signal $m(t)$, is non-linear. The application of the same test to the signal $r(t) = m(t) \cos(\omega_c t + \theta)$ shows that the process that generates $r(t)$ is linear.

The Fourier transform is a linear operator, that is, if a function can be written as a linear combination of other functions, the corresponding Fourier transform will be given by a linear combination of the corresponding Fourier transforms of each one of the functions involved in the linear combination (Gagliardi, 1988).

If $f(t) \longleftrightarrow F(\omega)$ and $g(t) \longleftrightarrow G(\omega)$, it then follows that

$$\alpha f(t) + \beta g(t) \longleftrightarrow \alpha F(\omega) + \beta G(\omega). \tag{1.74}$$

Proof: Let $h(t) = \alpha f(t) + \beta g(t) \rightarrow$, then it follows that

$$
\begin{aligned}
H(\omega) &= \int_{-\infty}^{\infty} h(t) e^{-j\omega t} dt \\
&= \alpha \int_{-\infty}^{\infty} f(t) e^{-j\omega t} dt + \beta \int_{-\infty}^{\infty} g(t) e^{-j\omega t} dt,
\end{aligned}
$$

and finally

$$H(\omega) = \alpha F(\omega) + \beta G(\omega). \tag{1.75}$$

1.6.2 Scaling Property

$$\mathcal{F}[f(at)] = \int_{-\infty}^{\infty} f(at) e^{-j\omega t} dt. \tag{1.76}$$

Initially, consider $a > 0$ in (1.76). By letting $u = at$, it follows that $dt = (1/a)du$. Replacing u for at in (1.76), one obtains

$$\mathcal{F}[f(at)] = \int_{-\infty}^{\infty} \frac{f(u)}{a} e^{-j\frac{\omega}{a} u} du$$

which simplifies to

$$\mathcal{F}[f(at)] = \frac{1}{a} F\left(\frac{\omega}{a}\right).$$

Consider the case in which $a < 0$. By a similar procedure, it follows that

$$\mathcal{F}[f(at)] = -\frac{1}{a} F\left(\frac{\omega}{a}\right).$$

Therefore, finally

$$\mathcal{F}[f(at)] = \frac{1}{|a|} F\left(\frac{\omega}{a}\right).\tag{1.77}$$

This result points to the fact that if a signal is compressed in the time domain by a factor a, then its Fourier transform will expand in the frequency domain by the same factor.

1.6.3 Symmetry of the Fourier Transform

This is an interesting property which can be fully observed in even functions. The symmetry property states that if

$$f(t) \longleftrightarrow F(\omega),\tag{1.78}$$

then it follows that

$$F(t) \longleftrightarrow 2\pi f(-\omega).\tag{1.79}$$

Proof: By definition,

$$f(t) = \frac{1}{2\pi} \int_{-\infty}^{+\infty} F(\omega)e^{j\omega t}d\omega,$$

which after multiplication of both sides by 2π becomes

$$2\pi f(t) = \int_{-\infty}^{+\infty} F(\omega)e^{j\omega t}d\omega.$$

By letting $u = -t$, it follows that

$$2\pi f(-u) = \int_{-\infty}^{+\infty} F(\omega)e^{-j\omega u}d\omega,$$

and now by making $t = \omega$, we obtain

$$2\pi f(-u) = \int_{-\infty}^{+\infty} F(t)e^{-jtu}dt.$$

Finally, by letting $u = \omega$, it follows that

$$2\pi f(-\omega) = \int_{-\infty}^{+\infty} F(t)e^{-j\omega t}dt.\tag{1.80}$$

Example: The Fourier transform of a constant function can be derived using the symmetry property. Since

$$A\delta(t) \longleftrightarrow A,$$

it follows that

$$A \longleftrightarrow 2\pi A\delta(-\omega) = 2\pi A\delta(\omega).$$

1.6.4 Time Domain Shift

Given that $f(t) \longleftrightarrow F(\omega)$, it then follows that $f(t - t_0) \longleftrightarrow F(\omega)e^{-j\omega t_0}$. Let $g(t) = f(t - t_0)$. In this case, it follows that

$$G(\omega) = \mathcal{F}[g(t)] = \int_{-\infty}^{\infty} f(t - t_0)e^{-j\omega t}dt. \tag{1.81}$$

By making $\tau = t - t_0$ it follows that

$$G(\omega) = \int_{-\infty}^{\infty} f(\tau)e^{-j\omega(\tau + t_0)}d\tau \tag{1.82}$$

$$= \int_{-\infty}^{\infty} f(\tau)e^{-j\omega\tau}e^{-j\omega t_0}d\tau, \tag{1.83}$$

and finally

$$G(\omega) = e^{-j\omega t_0}F(\omega). \tag{1.84}$$

This result shows that whenever a function is shifted in time, its frequency domain amplitude spectrum remains unaltered. However, the corresponding phase spectrum experiences a rotation proportional to ωt_0.

1.6.5 Frequency Domain Shift

Given that $f(t) \longleftrightarrow F(\omega)$, it then follows that $f(t)e^{j\omega_0 t} \longleftrightarrow F(\omega - \omega_0)$.

$$\mathcal{F}[f(t)e^{j\omega_0 t}] = \int_{-\infty}^{\infty} f(t)e^{j\omega_0 t}e^{-j\omega t}dt \tag{1.85}$$

$$= \int_{-\infty}^{\infty} f(t)e^{-j(\omega - \omega_0)t}dt,$$

$$\mathcal{F}[f(t)e^{j\omega_0 t}] = F(\omega - \omega_0). \tag{1.86}$$

1.6.6 Differentiation in the Time Domain

Given that

$$f(t) \longleftrightarrow F(\omega), \tag{1.87}$$

it then follows that

$$\frac{df(t)}{dt} \longleftrightarrow j\omega F(\omega). \tag{1.88}$$

Proof: Consider the expression for the inverse Fourier transform

$$f(t) = \frac{1}{2\pi} \int_{-\infty}^{\infty} F(\omega)e^{j\omega t}d\omega. \tag{1.89}$$

Differentiating in time, it follows that

$$\begin{aligned}
\frac{df(t)}{dt} &= \frac{1}{2\pi}\frac{\partial}{\partial t} \int_{-\infty}^{\infty} F(\omega)e^{j\omega t}d\omega \\
&= \frac{1}{2\pi} \int_{-\infty}^{\infty} F(\omega)\frac{\partial}{\partial t}e^{j\omega t}d\omega \\
&= \frac{1}{2\pi} \int_{-\infty}^{\infty} j\omega F(\omega)e^{j\omega t}d\omega,
\end{aligned}$$

and then

$$\frac{df(t)}{dt} \longleftrightarrow j\omega F(\omega). \tag{1.90}$$

In general, it follows that

$$\frac{d^n f(t)}{dt} \longleftrightarrow (j\omega)^n f(\omega). \tag{1.91}$$

By computing the Fourier transform of the signal $f(t) = \delta(t) - \alpha e^{-\alpha t}u(t)$, it is immediate to show that, by applying the property of differentiation in time, this signal is the time derivative of the signal $g(t) = e^{-\alpha t}u(t)$.

1.6.7 Integration in the Time Domain

Let $f(t)$ be a signal with zero average value, that is, let $\int_{-\infty}^{\infty} f(t)dt = 0$. By defining

$$g(t) = \int_{-\infty}^{t} f(\tau)d\tau, \tag{1.92}$$

it follows that

$$\frac{dg(t)}{dt} - f(t),$$

and since

$$g(t) \longleftrightarrow G(\omega), \tag{1.93}$$

then

$$f(t) \longleftrightarrow j\omega G(\omega),$$

and

$$G(\omega) = \frac{F(\omega)}{j\omega}. \tag{1.94}$$

In this manner, it follows that for a signal with zero average value

$$f(t) \longleftrightarrow F(\omega)$$

$$\int_{-\infty}^{t} f(\tau)d\tau \longleftrightarrow \frac{F(\omega)}{j\omega}. \tag{1.95}$$

Generalizing, for the case in which $f(t)$ has a non-zero average value, it follows that

$$\int_{-\infty}^{t} f(\tau)d\tau \longleftrightarrow \frac{F(\omega)}{j\omega} + \pi\delta(\omega)F(0). \tag{1.96}$$

1.6.8 The Convolution Theorem in the Time Domain

The convolution theorem can be used to analyze the frequency contents of a signal, to obtain many interesting results. One instance of the use of the convolution theorem, of fundamental importance in communication theory, is the sampling theorem which is discussed in the next section.

Let $h(t) = f(t) * g(t)$ and let $h(t) \longleftrightarrow H(\omega)$. It follows that

$$H(\omega) = \int_{-\infty}^{\infty} h(t)e^{-j\omega t}dt = \int_{-\infty}^{\infty}\int_{-\infty}^{\infty} f(\tau)g(t-\tau)e^{-j\omega t}dtd\tau. \tag{1.97}$$

$$H(\omega) = \int_{-\infty}^{\infty} f(\tau)\int_{-\infty}^{\infty} g(t-\tau)e^{-j\omega t}dtd\tau, \tag{1.98}$$

$$H(\omega) = \int_{-\infty}^{\infty} f(\tau)G(\omega)e^{-j\omega\tau}d\tau \tag{1.99}$$

and finally,

$$H(\omega) = F(\omega)G(\omega). \tag{1.100}$$

The convolution of two time functions is equivalent in the frequency domain to the product of their respective Fourier transforms.

1.6.9 The Convolution Theorem in the Frequency Domain

For the case in which a function in time is the product of two other functions, $h(t) = f(t) \cdot g(t)$, it is possible to obtain the Fourier transform proceeding in a way similar to the previous derivation.

$$H(\omega) = \frac{1}{2\pi} [F(\omega) * G(\omega)]. \tag{1.101}$$

The product of two time functions has a Fourier transform given by the convolution of their respective Fourier transforms. The convolution operation is often used when computing the response of a linear circuit, given its impulse response and an input signal.

Example: A given circuit has the impulse response $h(t)$ as follows,

$$h(t) = \frac{1}{RC} e^{-\frac{t}{RC}} u(t).$$

The application of the unit impulse $x(t) = \delta(t)$ as the input to this circuit causes an output $y(t) = h(t) * x(t)$. In the frequency domain, by the convolution theorem, it follows that $Y(\omega) = H(\omega)X(\omega) = H(\omega)$, that is, the Fourier transform of the impulse response of a linear system is the system transfer function.

Example: Using the frequency convolution theorem, it can be shown that

$$\cos(\omega_c t) u(t) \longleftrightarrow \frac{\pi}{2} [\delta(\omega + \omega_c) + \delta(\omega - \omega_c)] + j \frac{\omega}{\omega_c^2 - \omega^2}.$$

1.7 The Sampling Theorem

A band-limited signal $f(t)$, that has no frequency components above $\omega_M = 2\pi f_M$, can be reconstructed from its samples, collected at uniform time intervals $T_S = 1/f_S$, that is, at a sampling rate f_S, in which $f_S \geq 2f_M$. In fact, the condition for uniform time intervals is not necessary.

The sampling theory, derived by Claude E. Shannon, has been generalized for the case of non-uniform samples, that is, samples taken at non-equally spaced intervals (Davenport and Root, 1987).

It has been demonstrated that that a band-limited signal can be perfectly reconstructed from its samples, given that the average sampling rate satisfies the Nyquist condition, independent of the sampling being uniform or non-uniform. (Margolis and Eldar, 2008).

For a band-limited signal $f(t) \longleftrightarrow F(\omega)$, there is a frequency ω_M above which $F(\omega) = 0$, that is, that $F(\omega) = 0$ for $|\omega| > \omega_M$. Harry

Nyquist concluded that all the information about $f(t)$, shown in Figure 1.14, is contained in the samples of this signal, collected at regular time intervals T_S, as illustrated in Figure 1.15. In this way, the signal can be completely recovered from its samples.

For a band-limited signal $f(t)$, that is, such that $F(\omega) = 0$ for $|\omega| > \omega_M$, it follows that

$$f(t) * \frac{\sin(at)}{\pi t} = f(t), \text{ if } a > \omega_M,$$

because, in the frequency domain, this corresponds to the product of $F(\omega)$ by a gate function of width greater than $2\omega_M$.

The function $f(t)$ is sampled once every T_S seconds or, equivalently, sampled with a sampling frequency f_S, in which $f_S = 1/T_S \geq 2f_M$.

Consider the signal $f_s(t) = f(t)\delta_T(t)$, in which

$$\delta_T(t) = \sum_{n=-\infty}^{\infty} \delta(t - nT) \longleftrightarrow \omega_o \delta_{\omega_o} = \omega_o \sum_{n=-\infty}^{\infty} \delta(\omega - n\omega_o). \quad (1.102)$$

The periodic signal $\delta_T(t)$ is illustrated in Figure 1.16. The Fourier transform of the impulse train is represented in Figure 1.17.

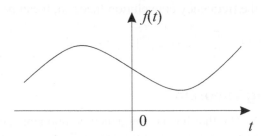

Figure 1.14 Band-limited signal $f(t)$.

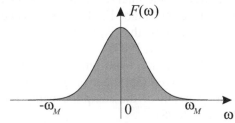

Figure 1.15 Spectrum of a band-limited signal.

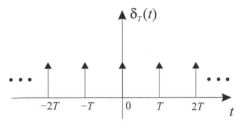

Figure 1.16 Impulse train used for sampling.

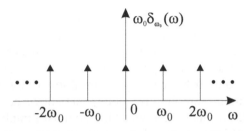

Figure 1.17 Fourier transform of the impulse train.

The signal $f_S(t)$ represents $f(t)$ sampled at uniform time intervals T_S seconds. From the frequency convolution theorem, it follows that the Fourier transform of the product of two functions in the time domain is given by the convolution of their respective Fourier transforms. It now follows that

$$f_S(t) \longleftrightarrow \frac{1}{2\pi}[F(\omega) * \omega_0 \delta_{\omega 0}(\omega)] \qquad (1.103)$$

The important frequency convolution theorem implies that the Fourier transform of the product of two functions in the time domain is given by the convolution of their respective Fourier transforms. It now follows that

$$f_S(t) \longleftrightarrow \frac{1}{2\pi}[F(\omega) * \omega_0 \delta_{\omega 0}(\omega)] \qquad (1.104)$$

and thus

$$f_S(t) \longleftrightarrow \frac{1}{T}[F(\omega) * \delta_{\omega 0}(\omega)] = \frac{1}{T} \sum_{n=-\infty}^{\infty} F(\omega - n\omega_o). \qquad (1.105)$$

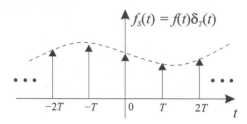

Figure 1.18   Example of a sampled signal.

From Figures 1.18 and 1.19, it can be observed that if the sampling frequency ω_S is less than $2\omega_M$, the spectral components will overlap. This will cause a loss of information because the original signal can no longer be completely recovered from its samples. As the signal frequency ω_S becomes smaller than $2\omega_M$, the sampling rate diminishes causing a partial loss of information.

Therefore, the minimum sampling frequency that allows perfect recovery of the signal is $\omega_S = 2\omega_M$, and is known as the Nyquist sampling rate, after Harry Nyquist (1889–1976), a Swedish engineer who made important contributions to communication theory. In order to recover the original spectrum $F(\omega)$, it is enough to pass the sampled signal through a low-pass filter with cut-off frequency ω_M.

For applications in digital telephony, the typical sampling frequency is $f_S = 8$ k samples/s. The speech signal is then quantized, using 256 distinct levels. Each level corresponds to an 8-bit binary code ($2^8 = 256$). After encoding, the signal is transmitted at a rate of 8 k samples/s $\times$ 8 bits/sample = 64 kbits/s, and the baseband signal occupies a bandwidth of approximately 64 kHz.

If the sampling frequency w_S is lower than $2\pi B$, in which B is the frequency in hertz, there will be spectra overlap and, as a consequence, information loss. As long as w_S becomes lower than $2\pi B$, the sampling rate becomes lower, leading to partial loss of information. Therefore, the sampling

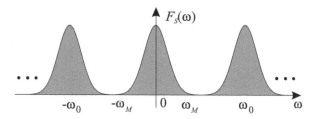

Figure 1.19   Spectrum of a sampled signal.

frequency for a baseband signal to be recovered without loss is $w_S = 2\pi B$, known as the Nyquist sampling frequency.

As mentioned, if the sampling frequency is lower than the Nyquist frequency, the signal will not be completely recovered, since there will be spectral superposition, leading to distortion in the highest frequencies. This phenomenon is known as aliasing, from the Latin word *alias*, meaning another, other, different.

On the other hand, increasing the sampling frequency above the Nyquist rate leads to spectra separation that is higher than the minimum necessary to recover the signal, causing a waste of spectrum usage in the transmission.

1.8 Parseval's Theorem

When dealing with a real signal $f(t)$ of finite energy, often called simply a real energy signal, the energy E associated with $f(t)$ is given by an integral in the time domain

$$E = \int_{-\infty}^{\infty} f^2(t)dt,$$

and can equivalently be calculated, in the frequency domain, by the formula

$$E = \frac{1}{2\pi} \int_{-\infty}^{\infty} |F(\omega)|^2 d.\omega$$

Equating both integrals, it follows that

$$\int_{-\infty}^{\infty} f^2(t)dt = \frac{1}{2\pi} \int_{-\infty}^{\infty} |F(\omega)|^2 d\omega. \tag{1.106}$$

The relationship given in (1.106) is known as Parseval's theorem or Parseval's identity. For a real signal $x(t)$ with energy E, it can be shown, by using Parseval's identity, that the signals $x(t)$ and its delayed version, $y(t) = x(t - \tau)$, have the same energy E.

Another way of expressing Parseval's identity is as follows.

$$\int_{-\infty}^{\infty} f(x)G(x)dx = \int_{-\infty}^{\infty} F(x)g(x)dx. \tag{1.107}$$

1.9 Average, Power, and Autocorrelation

As mentioned earlier, the average value of a real signal $x(t)$ is given by

$$\overline{x}(t) = \lim_{T \to \infty} \frac{1}{T} \int_{-\frac{T}{2}}^{\frac{T}{2}} x(t)dt. \tag{1.108}$$

The instantaneous power of $x(t)$ is given by

$$p_X(t) = x^2(t). \tag{1.109}$$

If the signal $x(t)$ exists for the whole interval $(-\infty, +\infty)$, the total average power $\overline{P}_X$ is defined for a real signal $x(t)$ as the power dissipated in a 1 ohm resistor, when a voltage $x(t)$ is applied to this resistor (or a current $x(t)$ flows through the resistor) (Lathi, 1989). Thus,

$$\overline{P}_X = \lim_{T \to \infty} \frac{1}{T} \int_{\frac{-T}{2}}^{\frac{T}{2}} x^2(t) dt \text{ W.} \tag{1.110}$$

From the previous definition, the unit to measure $\overline{P}_X$ corresponds to the square of the units of the signal $x(t)$ (either volt^2 or ampère^2, depending on the use of voltage or current). These units are commonly converted to watt, using a normalization by units of impedance (ohm).

It is common use to express the power in decibels (dBm), relative to the reference power of 1 mW. The power in dBm is given by the expression (Gagliardi, 1988)

$$\overline{P}_X = 10 \log \left[\frac{\overline{P}_X}{1 \text{ mW}} \right] \text{ dBm.} \tag{1.111}$$

The total power $(\overline{P}_X)$ contains two components: one constant component, because of the nonzero average value of the signal $x(t)$ $(\overline{P}_{DC})$, and an alternating component $(\overline{P}_{AC})$. The DC power of the signal is given by

$$\overline{P}_{DC} = [\overline{x}(t)]^2 \text{ W.} \tag{1.112}$$

Therefore, the AC power can be determined by removing the DC power from the total power, that is,

$$\overline{P}_{AC} = \overline{P}_X - \overline{P}_{DC} \text{ W.} \tag{1.113}$$

1.9.1 Time Autocorrelation of Signals

The average time autocorrelation $\overline{R}_X(\tau)$, or simply autocorrelation, of a real signal $x(t)$ is defined as follows

$$\overline{R}_X(\tau) = \lim_{T \to \infty} \frac{1}{T} \int_{\frac{-T}{2}}^{\frac{T}{2}} x(t) x(t + \tau) dt. \tag{1.114}$$

The change of variable $y = t + \tau$ allows Equation (1.114) to be written as

$$\overline{R}_X(\tau) = \lim_{T \to \infty} \frac{1}{T} \int_{\frac{-T}{2}}^{\frac{T}{2}} x(t) x(t - \tau) dt. \tag{1.115}$$

From Equations (1.114) and (1.115), it follows that $\overline{R}_X(\tau)$ is an even function of τ, and thus (Lathi, 1989)

$$\overline{R}_X(-\tau) = \overline{R}_X(\tau). \tag{1.116}$$

From the definition of autocorrelation and power, one obtains

$$\overline{P}_X = \overline{R}_X(0) \tag{1.117}$$

and

$$\overline{P}_{DC} = \overline{R}_X(\infty), \tag{1.118}$$

that is, from its autocorrelation function, it is possible to obtain information about the power of a signal. The AC power can be obtained as

$$P_{AC} = \overline{P}_X - \overline{P}_{DC} = \overline{R}_X(0) - \overline{R}_X(\infty). \tag{1.119}$$

The autocorrelation function can also be considered to obtain information about the function in the frequency domain by taking its Fourier transform, that is,

$$\mathcal{F}\{\overline{R}_X(\tau)\} = \int_{-\infty}^{+\infty} \lim_{T \to \infty} \frac{1}{T} \int_{-\frac{T}{2}}^{\frac{T}{2}} x(t)x(t+\tau)e^{-j\omega\tau} \, dt \, d\tau = \tag{1.120}$$

$$= \lim_{T \to \infty} \frac{1}{T} \int_{-\frac{T}{2}}^{\frac{T}{2}} x(t) \int_{-\infty}^{+\infty} x(t+\tau) \, d\tau \, dt$$

$$= \lim_{T \to \infty} \frac{1}{T} \int_{-\frac{T}{2}}^{\frac{T}{2}} x(t)X(\omega)e^{j\omega t} \, dt$$

$$= X(\omega) \lim_{T \to \infty} \frac{1}{T} \int_{-\frac{T}{2}}^{\frac{T}{2}} x(t)e^{j\omega t} \, dt$$

$$= \lim_{T \to \infty} \frac{X(\omega)X(-\omega)}{T}$$

$$= \lim_{T \to \infty} \frac{|X(\omega)|^2}{T} \tag{1.121}$$

The Power Spectral Density (PSD) $\overline{S}_X$ of a signal $x(t)$ is defined as the Fourier transform of the autocorrelation function $\overline{R}_X(\tau)$ of $x(t)$, that is, as

$$\overline{S}_X = \int_{-\infty}^{\infty} \overline{R}_X(\tau)e^{-j\omega\tau} \, d\tau. \tag{1.122}$$

Example: Find the PSD of the sinusoidal signal $x(t) = A\cos(\omega_0 t + \theta)$, illustrated in Figure 1.20.

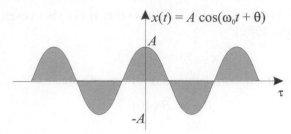

Figure 1.20   A sinusoidal signal with a deterministic phase.

Solution:

$$\overline{R}_X(\tau) = \lim_{T\to\infty} \frac{1}{T} \int_{\frac{-T}{2}}^{\frac{T}{2}} A^2 \cos(\omega_0 t + \theta) \cos\left[\omega_0(t + \tau) + \theta\right] dt$$

$$= \frac{A^2}{2} \lim_{T\to\infty} \frac{1}{T} \left[\int_{\frac{-T}{2}}^{\frac{T}{2}} \cos \omega_0 \tau dt + \int_{\frac{-T}{2}}^{\frac{T}{2}} \cos\left(2\omega_0 t + \omega_0 \tau + 2\theta\right) dt \right]$$

$$= \frac{A^2}{2} \cos \omega_0 \tau.$$

Notice that the autocorrelation function (Figure 1.21) is independent of the random phase θ. The power spectral density is given by

$$\overline{S}_X(\omega) = \mathcal{F}\left[R_X(\tau)\right]$$

$$\overline{S}_X(\omega) = \frac{\pi A^2}{2} \left[\delta(\omega + \omega_0) + \delta(\omega - \omega_0)\right],$$

which is illustrated in Figure 1.22.

The total power of the random signal $x(t)$ is computed making $\tau = 0$ in the autocorrelation function,

$$\overline{P}_X = \overline{R}_X(0) = \frac{A^2}{2}.$$

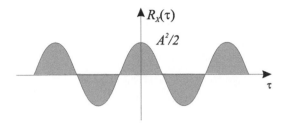

Figure 1.21 Autocorrelation function of a sinusoidal signal.

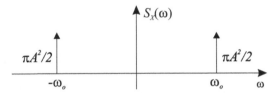

Figure 1.22 Power spectral density of a sinusoidal signal.

2

Random Signals and Noise

Deterministic signals are not found in real-life, therefore they are not always adequate to model real-world situations. The sinusoidal signal, for instance, is not realistic, because the signal source has to be switched on, and off, some time. On the other hand, circuit components, such as resistors, capacitors, inductors, and transistors, age and fail. Nature is usually random, and cause stochastic effects on communication systems. Noise is one of the best known phenomena to cause problems, disturb or disrupt signal transmission.

2.1 The Theory of Sets, Functions, and Measure

The axiomatic approach to the probability theory is based on the advanced set theory and on the measure theory. The advanced theory of sets was established by Georg Cantor (1845–1918) in the nineteenth century. Cantor set the basis for this theory and demonstrated some of its most important results, including the concept of set cardinality. Cantor was born in St. Petersburg, Russia, but lived most of his life in Germany (Boyer, 1974).

In 1872, J. W. R. Dedekind (1831–1916) pointed to the universal property of infinite sets, which has found applications as far as in the study of fractals (Boyer, 1974): A system is called infinite when it is similar to a part of itself. On the contrary, the system is finite.

Cantor also recognized the fundamental property of sets but, differing from Dedekind, he noticed that not all infinite sets are equal. This notion originated the cardinal numbers, which will be covered later, in order to establish a hierarchy of infinite sets in accordance with their respective powers. The results of Cantor led him to establish set theory as a fully developed subject (Alencar and Alencar, 2016).

2.1.1 Set Theory

The objective of this section is to develop the theory of sets in a summarized way, presenting the fundamental axioms. This theory is used as a basis to establish a probability measure, an important concept developed by the French mathematician Henry Lebesgue (1875–1941).

To illustrate the main idea of a set, some examples of common sets are given next.

- The binary set: $\mathbb{B} = \{0, 1\}$;
- The set of natural numbers: $\mathbb{N} = \{1, 2, 3, \dots\}$;
- The set of integer numbers: $\mathbb{Z} = \{\dots, -1, -2, 0, 1, 2, 3, \dots\}$;
- The set of real numbers: $\mathbb{R}$, which include the rational and irrational numbers.

The most important relations in set theory are the belonging relation, denoted as $a \in A$, in which a is an element of the set A, and the inclusion relation, $A \subset B$, which is read "A is a subset of the set B," or B is a superset of the set A.

Sets may be specified by means of propositions, as for example "The set of capitals", or more formally $A = \{a \mid a$ is a capital$\}$. The empty set can be written formally as $\emptyset = \{a \mid a \neq a\}$, that is, the set the elements of which are not equal to themselves.

A universal set is understood as that set which contains all other sets of interest. An example of a universal set is provided by the sample space in probability theory, usually denoted as S or Ω. The empty set is that set which contains no element and which is usually denoted as $\emptyset$ or $\{\ \}$. It is implicit that the empty set is contained in any set, this is, that $\emptyset \subset A$, for any given set A. However, the empty set is not in general an element of any other set.

Two sets are said to be disjoint if they have no element in common. Thus, for example, the set of even natural numbers and the set of odd natural numbers are disjoint.

2.1.2 Operations on Sets

- The operation $\overline{A}$ represents the complement of A with respect to the sample space Ω, that is, the set of all elements that do not belong to A.
- The subtraction of sets, denoted $C = A - B$, is the set of the elements that belong to A and do not belong to B. If B is completely contained in A, then $A - B = A \cap \overline{B}$.
- The set of elements that belong to A and to B, but do not belong to $A \cap B$ is specified by the symmetric difference $A \triangle B = A \cup B - A \cap B$.

The generalization of these concepts to families of sets, as for example $\cup_{i=1}^{N} A_i$ and $\cap_{i=1}^{N} A_i$, is immediate. The concepts are also useful when leading with the infinite, in countable algebras.

The following fundamental properties are usually employed in developing the theory of sets. They can simplify the deductions (Lipschutz, 1968).

- **Idempotent**

$$A \cup A = A, \qquad A \cap A = A \tag{2.1}$$

- **Associative**

$$(A \cup B) \cup C = A \cup (B \cup C) \tag{2.2}$$
$$(A \cap B) \cap C = A \cap (B \cap C) \tag{2.3}$$

- **Commutative**

$$A \cup B = B \cup A \tag{2.4}$$
$$A \cap B = B \cap A \tag{2.5}$$

- **Distributive**

$$A \cup (B \cap C) = (A \cup B) \cap (A \cup C) \tag{2.6}$$
$$A \cap (B \cup C) = (A \cap B) \cup (A \cap C) \tag{2.7}$$

- **Identity**

$$A \cup \emptyset = A, \qquad A \cap U = A \tag{2.8}$$
$$A \cup U = U, \qquad A \cap \emptyset = \emptyset \tag{2.9}$$

- **Complementary**

$$A \cup \overline{A} = U, \qquad A \cap \overline{A} = \emptyset, \qquad \overline{(\overline{A})} = A \tag{2.10}$$
$$\overline{U} = \emptyset, \qquad \overline{\emptyset} = U \tag{2.11}$$

- **de Morgan laws**

$$\overline{A \cup B} = \overline{A} \cap \overline{B} \tag{2.12}$$
$$\overline{A \cap B} = \overline{A} \cup \overline{B} \tag{2.13}$$

2.1.3 Families of Sets

A family of sets is a collection of subsets of a given set. This definition of family of sets allows repeated elements, that characterizes a multi-set, but the elements must be distinguishable. For example, $\mathcal{F} = \{A_1, A_2, A_3, A_4, A_5\}$, with $A_1 = \{a, b\}$, $A_2 = \{a, b, c, d\}$, $A_3 = \{a, b, c, d, e\}$, $A_4 = \{a, b, c, d\}$, and $A_5 = \{a, b\}$, is a family of sets.

Among the most interesting families of sets, it is worthy to mention an increasing sequence of sets, such that $\lim_{i \to \infty} \cup A_i = A$. This sequence is used in proofs of limits over sets. A decreasing sequence of sets is defined in a similar way, with $\lim_{i \to \infty} \cap A_i = A$.

2.1.4 Indexing Sets

The Cartesian product is a convenient way to express the idea of set indexing. The operation of set indexing, the association of a label to each set in the family, expands the possibilities for the use of sets, and permits the production of vectors and signals, for example, which are useful concepts in Mathematics, Physics, and Engineering.

Example: Consider the set $A_i = \{0, 1\}$. Starting from this set, it is possible to construct an indexed sequence of sets by defining its indexing: $\{A_{i \in I}\}$, $I = \{0, \cdots, 7\}$. This family of indexed sets A_i constitutes a finite discrete sequence, that is, a vector. For example, the American Standard Code for Information Interchange (ASCII) can be represented in this way.

Example: Consider again $A_i = \{0, 1\}$, and let $I = Z$, the set of positive and negative integers plus zero. It follows that $\{A_{i \in Z}\}$, which represents an infinite series of 0's and 1's, that is, it represents a binary digital signal, such as $\cdots 101010011010 \cdots$.

Example: As another example, consider $A_i = \{0, 1\}$, but let the indexing be over the set of real numbers, $\{A_{i \in I}\}$, in which $I = R$, a signal is formed which is discrete in amplitude but continuous in time, such as the telegraph signal sketched in Figure 2.1.

Example: An analog signal, that is continuous in time and in amplitude, is modeled, using families of sets, as $A = R$ and $I = R$. This is shown in Figure 2.2.

2.1.5 Algebra of Sets

In order to construct an algebra of sets or, equivalently, to construct a field over which operations involving sets make sense, a few properties have to be obeyed.

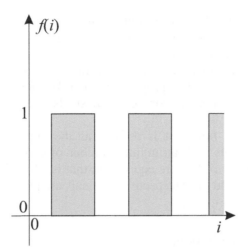

Figure 2.1 A telegraphic signal that is discrete in amplitude and continuous in time.

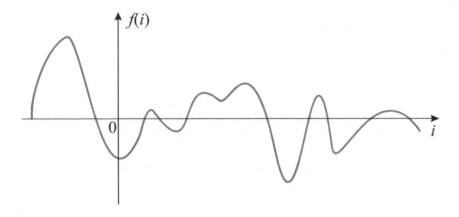

Figure 2.2 An analog signal that is both continuous in time and amplitude.

1. If $A \in \mathcal{F}$, then $\overline{A} \in \mathcal{F}$. A is the set containing desired results, or over which one wants to operate.
2. If $A \in \mathcal{F}$ and $B \in \mathcal{F}$, then $A \cup B \in \mathcal{F}$.

The above properties guarantee the closure of the algebra with respect to finite operations over sets. It is noticed that the universal set Ω always belongs to the algebra, that is, $\Omega \in \mathcal{F}$, because $\Omega = A \cup \overline{A}$. The empty set also belongs to the algebra, that is, $\emptyset \in \mathcal{F}$, since $\emptyset = \overline{\Omega}$, follows by property 1.

Example: The family $\{\emptyset, \Omega\}$ complies with the above properties and therefore represents an algebra. In this case, $\emptyset = \{\}$ and $\bar{\emptyset} = \Omega$. The union is also represented, as can be easily checked.

If there is a measure for heads, then there must also be a measure for tails, in order for the algebra to be properly defined. Whenever a probability is assigned to an event then a probability must also be assigned to the complementary event.

The cardinality of a finite set is defined as the number of elements belonging to this set. Sets with an infinite number of elements are said to have the same cardinality if they are equivalent, that is, $A \sim B$ if $\sharp A = \sharp B$. Some examples of sets and their respective cardinals are presented next.

- $I = \{1, \cdots, k\} \Rightarrow C_I = k$;

- $N = \{0, 1, \cdots\} \Rightarrow C_N$ or $\aleph_0$;

- $Z = \{\cdots, -2, -1, 0, 1, 2, \cdots\} \Rightarrow C_Z$;

- $Q = \{\cdots, -1/3, 0, 1/3, 1/2, \cdots\} \Rightarrow C_Q$;

- $R = (-\infty, \infty) \Rightarrow C_R$ or $\aleph$.

The following relations involving the cardinalities of the previous sets can be verified: $C_R > C_Q = C_Z = C_N > C_I$. The notation $\aleph_0$ for the cardinality of the set of natural numbers, which uses the first letter of the Hebrew alphabet, was introduced by Cantor.

The cardinality of the power set, that is, of the family of sets consisting of all subsets of a given set A, $\mathcal{F} = 2^A$, is 2^{C_A}.

2.1.6 Borel Algebra

The Borel algebra $\mathcal{B}$, or σ-algebra, is an extension of the algebra so far discussed to operate with limits at infinity. The following properties are required from a σ-algebra.

1. $A \in \mathcal{B} \Rightarrow \overline{A} \in \mathcal{B}$;
2. $A_i \in \mathcal{B} \Rightarrow \bigcup_{i=1}^{\infty} A_i \in \mathcal{B}$.

The above properties guarantee the closure of the σ-algebra with respect to enumerable operations over sets. These properties allow the definition of limits in the Borel field.

Example: Considering the above properties, it can be verified that $A_1 \cap A_2 \cap A_3 \cdots \in \mathcal{B}$. In effect, it is sufficient to notice that

$$A \in \mathcal{B} \text{ and } \mathcal{B} \in \mathcal{B} \Rightarrow A \cup \mathcal{B} \in \mathcal{B},$$

and

$$\overline{A} \in \mathcal{B} \text{ and } \overline{\mathcal{B}} \in \mathcal{B} \Rightarrow \overline{A} \cup \overline{\mathcal{B}} \in \mathcal{B},$$

and finally

$$\overline{\overline{A} \cup \overline{\mathcal{B}}} \in \mathcal{B} \Rightarrow A \cap \mathcal{B} \in \mathcal{B}.$$

In summary, any combination of unions and intersections of sets belongs to the Borel algebra. In other words, operations of union or intersection of sets, or a combination of these operations, produce a set that belongs to the σ-algebra.

2.2 Probability Theory

This section presents the basic definitions related to the theory of probability, the main results and conclusions of which will be used in subsequent sections and chapters.

2.2.1 Axiomatic Approach to Probability

The axioms of probability were established by Andrei N. Kolmogorov (1903–1987), allowing the development of the complete theory. There are just three statements, as follows (Papoulis, 1983a):

Axiom 1 – $P(\Omega) = 1$, in which Ω denotes the sample space or universal set and $P(\cdot)$ denotes the associated probability.

Axiom 2 – $P(A) \geq 0$, in which A denotes an event belonging to the sample space.

Axiom 3 – $P(A \cup B) = P(A) + P(B)$, in which A and B are mutually exclusive events and $A \cup B$ denotes the union of events A and B.

Kolmogorov's fundamental work on probability was published in 1933, in Russian, and later on it was published in German, with the title *Grundbegriffe der Wahrscheinlichkeits Rechnung* (Fundamentals of Probability Theory) (James, 1981). In this work, Kolmogorov managed to combine Advanced Set Theory, of Cantor, with Measure Theory, of Lebesgue, in order to produce what to this date is the modern approach to probability theory.

The application of the axioms makes it possible to deduce all results of probability theory. For example, the probability of the empty set, $\emptyset = \{\}$, is calculated as follows. First, it is noticed that

$$\emptyset \cup \Omega = \Omega,$$

since the sets $\emptyset$ and Ω are disjoint. Thus, it follows that

$$P(\emptyset \cup \Omega) = P(\Omega) = P(\emptyset) + P(\Omega) = 1 \Rightarrow P(\emptyset) = 0.$$

In the case of sets A and B which are not disjoint, it follows that

$$P(A \cup B) = P(A) + P(B) - P(A \cap B). \tag{2.14}$$

2.2.2 Bayes' Rule

Bayes' rule concerns the computation of conditional probabilities, which are the basis of information theory, and can be expressed by the following rule

$$P(A|B) = \frac{P(A \cap B)}{P(B)}, \tag{2.15}$$

assuming $P(B) \neq 0$.

An equivalent manner to express the same result is the following,

$$P(A \cap B) = P(A|B) \cdot P(B) , \quad P(B) \neq 0. \tag{2.16}$$

Some important properties of sets are presented next, in which A and B denote events from a given sample space.

- If A is independent of B, then $P(A|B) = P(A)$. It then follows that $P(B|A) = P(B)$ and that B is independent of A.
- If $B \subset A$, then: $P(A|B) = 1$.
- If $A \subset B$, then: $P(A|B) = \frac{P(A)}{P(B)} \geq P(A)$.
- If A and B are independent events, then $P(A \cap B) = P(A) \cdot P(B)$.
- If $P(A) = 0$ or $P(A) = 1$, then event A is independent of itself.
- If $P(B) = 0$, then $P(A|B)$ can assume any arbitrary value. Usually, in this case, one assumes $P(A|B) = P(A)$.
- If events A and B are disjoint, and non-empty, then they are certainly dependent.

A partition is a possible splitting of the sample space into a family of subsets, in such a way that the subsets in this family are disjoint and their union coincides with the sample space.

It follows that any set in the sample space can be expressed using a partition of that space, and thus can be written as a union of disjoint events. This is a useful property when computing probabilities.

The partition property states that if a set can be composed of the intersection of the original set and a partition of the universal set

$$B = B \cap \Omega = B \cap \cup_{i=1}^{M} A_i = \cup_{i=1}^{N} B \cap A_i,$$

then, it follows that

$$P(B) = P(\cup_{i=1}^{N} B \cap A_i) = \sum_{i=1}^{N} P(B \cap A_i), \tag{2.17}$$

and

$$
\begin{aligned}
P(A_i|B) &= \frac{P(A_i \cap B)}{P(B)} = \frac{P(B|A_i) \cdot P(A_i)}{\sum_{i=1}^{N} P(B \cap A_i)} \\
&= \frac{P(B|A_i) \cdot P(A_i)}{\sum_{i=1}^{N} P(B|A_i) \cdot P(A_i)}.
\end{aligned} \tag{2.18}
$$

2.3 Random Variables

A random variable X represents a mapping of the sample space on the line (the set of real numbers). A random variable is usually characterized by a cumulative probability function (CPF) $P_X(x)$, or by a probability density function (pdf) $p_X(x)$.

Example: A random variable with a uniform pdf $p_X(x)$ is described by the equation $p_X(x) = u(x) - u(x - 1)$. It follows by Axiom 1 that

$$\int_{-\infty}^{+\infty} p_X(x) dx = 1. \tag{2.19}$$

For a given probability distribution, the probability that the random variable X belongs to the interval $(a, b]$ is given by

$$P(a < x \le b) = \int_{a}^{b} p_X(x) dx. \tag{2.20}$$

The cumulative density function (CDF) $P_X(x)$, of a random variable X, is defined as the integral of the pdf $p_X(x)$, that is,

$$P_X(x) = \int_{-\infty}^{r} p_X(t) dt. \tag{2.21}$$

2.3.1 Mean Value of a Random Variable

Most of the results in communication theory are derived from the computation of averages. Let $f(X)$ denote a function of a random variable X. The average value (or expected value) of $f(X)$ with respect to X is defined as

$$E[f(X)] = \int_{-\infty}^{+\infty} f(x)p_X(x)dx. \tag{2.22}$$

The following properties of the expected value follow from (2.22).

$$E[\alpha X] = \alpha E[X], \tag{2.23}$$
$$E[X + Y] = E[X] + E[Y] \tag{2.24}$$

and if X and Y are independent random variables, then

$$E[XY] = E[X]E[Y]. \tag{2.25}$$

2.3.2 Moments of a Random Variable

A moment is an extracted feature of the distribution, which reveals a certain characteristic of the random variable. The k-th moment of a random variable X is defined as

$$m_k = E[X^k] = \int_{-\infty}^{+\infty} x^k p_X(x)dx. \tag{2.26}$$

Some moments of X have special importance and present interesting physical interpretations.

1. $m_1 = E[X]$ is the arithmetic mean, also called average value, average voltage, or statistical mean of the random variable;
2. $m_2 = E[X^2]$ represents the quadratic mean or total power of the signal;
3. $m_3 = E[X^3]$ is a measure of asymmetry of the pdf, and reveals the tendency of the random variable;
4. $m_4 = E[X^4]$ is a measure of flatness of the pdf. It is usually compared to the Gaussian random variable.

2.3.3 The Variance of a Random Variable

The variance of a random variable X is an important quantity in communication theory and electronics, because it represents the varying signal power, defined as follows.

$$V[X] = \sigma_X^2 = E[(X - m_1)^2] = m_2 - m_1^2. \tag{2.27}$$

The standard deviation σ_X is defined as the square root of the variance of the random variable X.

2.3.4 The Characteristic Function of a Random Variable

The characteristic function $P_X(w)$, or moment generating function, of a random variable X is usually defined from the Fourier transform of the pdf of X, which is equivalent to making $f(x) = e^{-j\omega x}$ in Formula 2.22, that is,

$$P_X(w) = \mathrm{E}[e^{-j\omega x}] = \int_{-\infty}^{+\infty} e^{-j\omega x} p_X(x)dx, \text{ in which } j = \sqrt{-1}. \quad (2.28)$$

The moments of a random variable X can also be obtained directly from then characteristic function as follows.

$$m_i = \frac{1}{(-j)^i} \frac{\partial^i P_X(w)}{\partial w^i}\Big|_{w=0}. \quad (2.29)$$

Given that X is a random variable, it follows that $Y = f(X)$ is also a random variable, obtained by the application of the transformation $f(\cdot)$. The pdf of Y is related to that of X by the formula (Blake, 1987)

$$p_Y(y) = \frac{p_X(x)}{|dy/dx|}\Big|_{x=f^{-1}(y)}, \quad (2.30)$$

in which $f^{-1}(\cdot)$ denotes the inverse function of $f(\cdot)$. This formula assumes the existence of the inverse function of $f(\cdot)$ as well as its derivative in all points.

2.3.5 Some Important Random Variables

Some random variables are particularly important to electrical engineering, because they model physical phenomena. That is the case with the Gaussian, Rayleigh, and sinusoidal random variables.

1. **Gaussian random variable**
   The random variable X with pdf

$$p_X(x) = \frac{1}{\sigma_X \sqrt{2\pi}} e^{-\frac{(x-m_X)^2}{2\sigma_X^2}} \quad (2.31)$$

is called a Gaussian (or Normal) random variable. The Gaussian random variable plays an extremely important role in engineering, considering

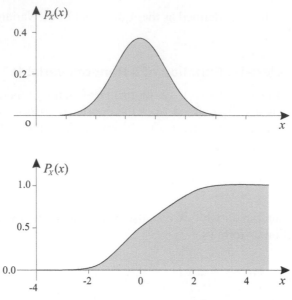

Figure 2.3   Gaussian probability density and corresponding cumulative probability function.

that many well-known processes can be described or approximated by this pdf.

The noise present in either analog or digital communication systems usually can be considered Gaussian as a consequence of the influence of many factors (Leon-Garcia, 1989). In (2.31), m_X represents the average value and σ_X^2 represents the variance of X. Figure 2.3 illustrates the Gaussian pdf and its corresponding CDF.

2. **Rayleigh random variable**
   An often used model to represent the behavior of the amplitudes of signals subjected to fading employs the following pdf (Kennedy, 1969).

$$p_X(x) = \frac{x}{\sigma^2} e^{-\frac{x^2}{2\sigma^2}} u(x) \tag{2.32}$$

known as the Rayleigh pdf, with average $E[X] = \sigma\sqrt{\pi/2}$ and variance $V[X] = (2 - \pi)\frac{\sigma^2}{2}$ (Proakis, 1990).

The Rayleigh pdf represents the effect of multiple signals, reflected or refracted, which are captured by a receiver, in a situation in which there is no main signal component or main direction of propagation (Lecours et al., 1988). In this situation, the phase distribution of the received signal can be considered uniform in the interval $(0, 2\pi)$.

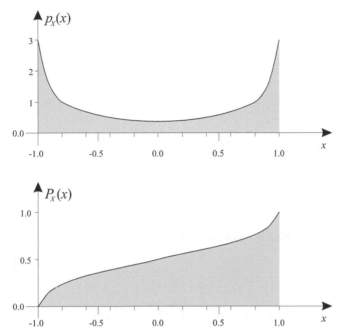

Figure 2.4 Probability density function and cumulative probability function of a sinusoidal random variable.

Researchers have found that it is possible to closely approximate a Rayleigh pdf by considering only six waveforms with independently distributed phases (Schwartz et al., 1966).

3. **Sinusoidal random variable**

A sinusoidal tone X has the following pdf

$$p_X(x) = \frac{1}{\pi\sqrt{V^2 - x^2}}, \ |x| < V. \tag{2.33}$$

The pdf and the CPF of X are illustrated in Figure 2.4.

2.3.6 Joint Random Variables

Considering that X and Y represent a pair of real random variables, with joint pdf $p_{XY}(x, y)$, as illustrated in Figure 2.5, then the probability of x and y being simultaneously in the region defined by the polygon [abcd] is given by the expression

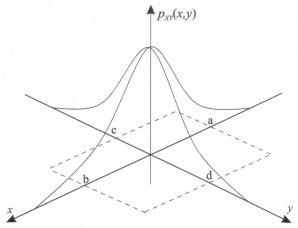

Figure 2.5   Joint probability density function.

$$\text{Prob}(a < x < b, c < y < d) = \int_a^b \int_c^d p_{XY}(x,y)dxdy. \qquad (2.34)$$

The individual pdfs of X and Y, also called marginal pdfs, result from the integration of the joint pdf as follows.

$$p_X(x) = \int_{-\infty}^{+\infty} p_{XY}(x,y)dy, \qquad (2.35)$$

and

$$p_Y(y) = \int_{-\infty}^{+\infty} p_{XY}(x,y)dx. \qquad (2.36)$$

The joint average $\mathrm{E}[f(X,Y)]$ is calculated as

$$\mathrm{E}[f(X,Y)] = \int_{-\infty}^{+\infty} \int_{-\infty}^{+\infty} f(x,y)p_{XY}(x,y)dxdy, \qquad (2.37)$$

for an arbitrary function $f(X,Y)$ of X and Y.

The joint moments m_{ik}, of order ik, are calculated as

$$m_{ik} = \mathrm{E}[X^i, Y^k] = \int_{-\infty}^{+\infty} \int_{-\infty}^{+\infty} x^i y^k p_{XY}(xy)dxdy. \qquad (2.38)$$

The two-dimensional characteristic function is defined as the two-dimensional Fourier transform of the joint probability density $p_{XY}(x, y)$

$$P_{XY}(\omega, \nu) = \mathrm{E}[e^{-j\omega X - j\nu Y}]. \qquad (2.39)$$

When the sum $Z = X + Y$ of two statistically independent random variables is considered, it is noticed that the characteristic function of Z turns out to be

$$P_Z(\omega) = \mathrm{E}[e^{-j\omega Z}] = \mathrm{E}[e^{-j\omega(X+Y)}] = P_X(\omega) \cdot P_Y(\omega). \qquad (2.40)$$

As far as the pdf of Z is concerned, it can be said that

$$p_Z(z) = \int_{-\infty}^{\infty} p_X(\rho)p_Y(z - \rho)d\rho, \qquad (2.41)$$

or

$$p_Z(z) = \int_{-\infty}^{\infty} p_X(z - \rho)p_Y(\rho)d\rho. \qquad (2.42)$$

Equivalently, the sum of two statistically independent random variables has a pdf given by the convolution of the respective pdfs of the random variables involved in the sum.

The random variables X and Y are uncorrelated if $\mathrm{E}[XY] = \mathrm{E}[X]\mathrm{E}[Y]$. The criterion of statistical independence of random variables, which is stronger than that for the random variables being uncorrelated, is satisfied if $p_{XY}(x, y) = p_X(x).p_Y(y)$.

2.4 Stochastic Processes

A random signal, also known generically as a stochastic process, is an extension of the concept of a random variable, involving a time variable. In fact, a stochastic process is a function of a random variable, or of a combination of random variables, and time. Figure 2.6 illustrates a random signal and its associated pdf.

A random process $X(t)$ defines a random variable for each point on the time axis. A stochastic process is said to be stationary if the probability densities associated with the process do not change with time.

2.4.1 The Autocorrelation Function

The autocorrelation function is an important joint moment of the random process $X(t)$, because it reveals certain characteristics of the signal, that could not be apparent at first sight. It is used to detect signals in noise, but has many other applications.

$$R_X(\xi, \eta) = \mathrm{E}[X(\xi)X(\eta)], \tag{2.43}$$

in which,

$$\mathrm{E}[X(\xi)X(\eta)] = \int_{-\infty}^{+\infty} \int_{-\infty}^{+\infty} x(\xi)x(\eta)p_{X(\xi)X(\eta)}(x(\xi)x(\eta))dx(\xi)dy(\eta) \tag{2.44}$$

denotes the joint moment of the random variable. $X(t)$ at $t = \xi$ and at $t = \eta$.

The random process is called wide sense stationary if its autocorrelation depends only on the interval of time separating $X(\xi)$ and $X(\eta)$, that is, depends only on $\tau = \xi - \eta$. Equation (2.43) in this case can be written as

$$R_X(\tau) = \mathrm{E}[X(t)X(t + \tau)]. \tag{2.45}$$

2.4.2 Stationarity

In general, the statistical mean of a random signal is a function of time, that is, it changes with time. Thus, the mean value

$$\mathrm{E}[X(t)] = m_X(t),$$

the power

$$\mathrm{E}[X^2(t)] = P_X(t)$$

and the autocorrelation

$$R_X(\tau, t) = \mathrm{E}[X(t)X(t + \tau)],$$

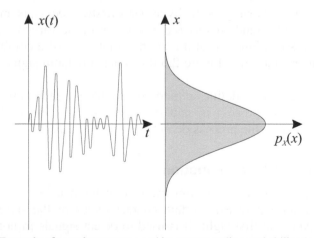

Figure 2.6 Example of a random process and its corresponding probability density function.

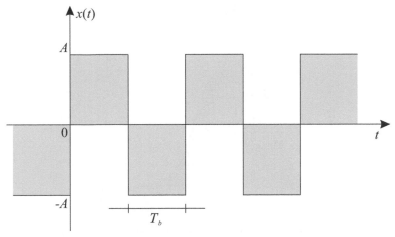

Figure 2.7 A digital signal.

are, in general, time dependent. However, there exists a set of time signals the mean value of which is time independent. These signals are called stationary signals, as illustrated in the following example.

Example: Consider the digital signal shown in Figure 2.7, with equiprobable amplitudes A and $-A$.

The pdf of the digital signal, as shown in Figure 2.8, is given by

$$p_X(x) = \frac{1}{2}[\delta(x + A) + \delta(x - A)].$$

Applying the definition of the mean to $E[X(t)]$, it follows that

$$E[X(t)] = 0.$$

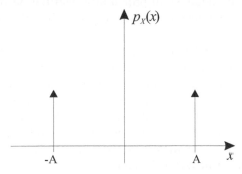

Figure 2.8 Probability density function for the digital signal of Figure 2.7.

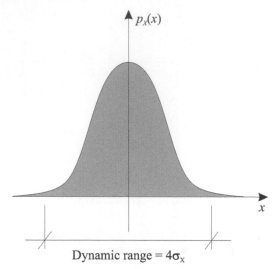

Figure 2.9   The dynamic range of a signal.

The power is given as,

$$E[X^2(t)] = \frac{1}{2}A^2 + \frac{1}{2}(-A)^2 = A^2.$$

Finally, the variance and the standard deviation are calculated as

$$\sigma_X^2 = E[X^2(t)] = A^2 \Rightarrow \sigma_X = A.$$

Example: The dynamic range of a signal, from a probabilistic point of view, is illustrated in Figure 2.9. As can be seen, the dynamic range depends on the standard deviation, or RMS voltage, being usually specified for $2\sigma_X$ or $4\sigma_X$. For a signal with a Gaussian probability distribution of amplitudes, this corresponds to a range encompassing, respectively, 97% and 99.7% of all signal amplitudes.

However, since the signal is time varying, its statistical mean can also change with time, as illustrated in Figure 2.10. In the example considered, the variance is initially diminishing with time and later it is growing with time. In this case, an adjustment in the signal variance, by means of an automatic gain control mechanism, can remove the pdf dependency on time.

A signal is stationary whenever its pdf is time independent, that is, whenever $p_X(x,t) = p_X(x)$, as illustrated in Figure 2.11.

Stationarity may occur in various instances:

1. The signal is stationary in the mean when the mean is independent of time, $m_X(t) = m_X$.

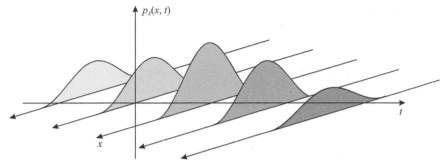

Figure 2.10 A time-varying probability density function.

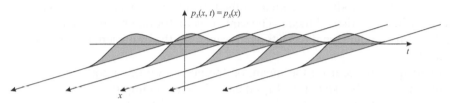

Figure 2.11 A time-invariant probability density function.

2. The signal is stationary in power when the power is independent of time, $P_X(t) = P_X$.
3. First-order stationarity implies that the first-order moment is indepen-dent of time.
4. Second-order stationarity implies that the first- and second-order moments of the signal are independent of time.
5. Strict sense stationarity implies that the signal is stationary for all orders, that is, $p_{X_1 \cdots X_M}(x_1, \cdots, x_M; t) = p_{X_1 \cdots X_M}(x_1, \cdots, x_M)$.

A real stochastic signal is seldom strictly stationary, for the whole time. But, most signals can be considered stationary, in the mean and power, if some conditions are met. For instance, if the time interval is restricted, or if the audio equipment or the communication receiver has automatic gain control.

2.4.3 Wide Sense Stationarity

However, there is a type of stationarity that is useful to most purposes, but does not require the signal to fulfill many statistical tests. The following conditions are necessary to guarantee that a stochastic process is wide sense stationary.

1. The signal mean and the power are constant.
2. The autocorrelation depends only on the time difference between the measurements,

$$R_X(t_1, t_2) = R_X(t_2 - t_1) = R_X(\tau),$$

that is, the autocorrelation does not depend on the origin of the time interval.

2.4.4 Ergodic Signals

Ergodicity is another important characteristic of random processes. For simplicity, a signal is ergodic in the mean if the time expected value of the signal coincides with its statistical mean. The same is valid for other signal statistics. Therefore, an ergodic signal has a typical waveform that represents the whole ensemble.

Ergodicity can occur on the mean, on the power, on the autocorrelation, or with respect to other statistics. Ergodicity of the mean implies that the time average is equivalent to the statistical signal average. Therefore,

1. Ergodicity of the mean – $\overline{X(t)} \sim E[X(t)]$.
2. Ergodicity of the power – $\overline{X^2(t)} \sim \overline{R_X(\tau)} \sim R_X(\tau)$.
3. Ergodicity of the autocorrelation – $\overline{R_X(\tau)} \sim R_X(\tau)$.

A strictly stationary stochastic process has a joint pdf that is independent of time. A wide sense stationary process has the first- and second-order means constant and the autocorrelation depending only on the measuring time interval.

Therefore, a stochastic process is ergodic whenever its statistical means, which are functions of time, can be approximated by their corresponding time averages, which are random processes, with a standard deviation which is close to zero. The ergodicity may appear only on the mean value of the process, in which case the process is said to be ergodic in the mean.

2.4.5 Properties of the Autocorrelation

It is not necessary to compute the autocorrelation using the definition every time. The autocorrelation function has some important properties, presented in the following, that help solve most of the problems.

1. The total power of a signal is obtained by computing the autocorrelation at the origin,

$$R_X(0) = E[X^2(t)] = P_X. \tag{2.46}$$

2. The average power or DC power level is the limit of the autocorrelation, as the time difference goes to infinity,

$$R_X(\infty) = \lim_{\tau \to \infty} R_X(\tau) = \lim_{\tau \to \infty} \mathrm{E}[X(t + \tau)X(t)] = \mathrm{E}^2[X(t)].$$

(2.47)

3. The signal mean value is the square root of the autocorrelation evaluated at the infinity,

$$\mathrm{E}[X(t)] = \sqrt{R_X(\infty)}.$$

(2.48)

4. The autocovariance is the unbiased autocorrelation,

$$C_X(\tau) = \mathrm{E}[(X(t + \tau) - m_X)(X(t) - m_X)]$$
$$= R_X(\tau) - \mathrm{E}^2[X(t)].$$

(2.49)

5. The variance, or AC power, is the unbiased total power,

$$V[X(t)] = \mathrm{E}[(X(t) - \mathrm{E}[X(t)])^2]$$
$$= \mathrm{E}[X^2(t)] - \mathrm{E}^2[X(t)]$$
$$= R_X(0) - R_X(\infty) = P_{AC}.$$

(2.50)

6. The autocorrelation has a maximum at the origin,

$$R_X(0) \geq |R_X(\tau)|.$$

(2.51)

This property is demonstrated by considering the following mathematical tautology, that is always true for real signals,

$$\mathrm{E}[(X(t) - X(t + \tau))^2] \geq 0.$$

Thus,

$$\mathrm{E}[X^2(t) - 2X(t)X(t + \tau)] + \mathrm{E}[X^2(t + \tau)] \geq 0,$$

that is,

$$2R_X(0) - 2RX(\tau) \geq 0 \ \Rightarrow \ R_X(0) \geq R_X(\tau).$$

7. The autocorrelation is a symmetric function,

$$R_X(\tau) = R_X(-\tau).$$

(2.52)

In order to prove this property, it is sufficient to use the definition

$$R_X(-\tau) = \mathrm{E}[X(t)X(t - \tau)].$$

Considering a change of variables, $t - \tau = \sigma \ \Rightarrow \ t = \sigma + \tau$, one obtains

$$R_X(-\tau) = \mathrm{E}[X(\sigma + \tau) \cdot X(\sigma)] = R_X(\tau).$$

The relationship between the autocorrelation function and various other power measures is illustrated in Figure 2.12.

Example: The autocorrelation of a digital signal $X(t)$, with equiprobable amplitude levels A and $-A$, and a random initial transition, as seen in Figure 2.13, with uniform distribution, can be computed in the following manner.

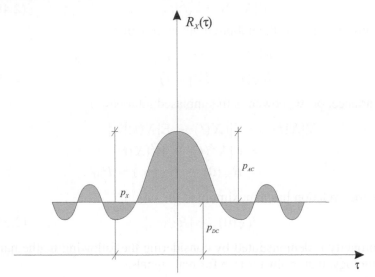

Figure 2.12   Relationship between the autocorrelation and various power measures.

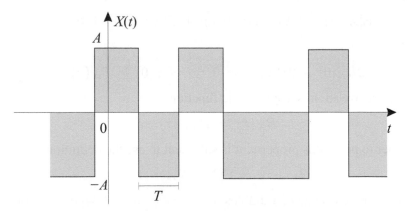

Figure 2.13   Digital random signal.

A baseband random binary signal assumes only two states, $-A$ and A, and the transition between the states can take place every T_b seconds. The autocorrelation can be computed by (Lathi, 1968)

$$R_X(\tau) = \sum_{X_1} \sum_{X_2} x_1 x_2 P_{X_1 X_2}(x_1, x_2) \tag{2.53}$$

in which x_1 and x_2 represent the signal amplitudes at t and $t + \tau$, and the summations are over the sets $X_1 = X_2 = \{-A, A\}$. This can be written as

$$\begin{aligned} R_X(\tau) = {} & A^2 \left[P_{X_1 X_2}(A, A) + P_{X_1 X_2}(-A, -A) \right. \\ & \left. - P_{X_1 X_2}(-A, A) - P_{X_1 X_2}(A, -A) \right]. \end{aligned} \tag{2.54}$$

But,

$$P_{X_1 X_2}(A, A) = P_{X_1}(A) P_{X_2}(A|x_1 = A),$$

and, since the signal states are equiprobable,

$$P_{X_1}(A) = P_{X_1}(-A) = 0.5.$$

Therefore, it is possible to write

$$\begin{aligned} R_X(\tau) = {} & A^2 P_{X_1}(A) P_{X_2}(A|x_1 = A) \\ & + A^2 P_{X_1}(-A) P_{X_2}(-A|x_1 = -A) \\ & - A^2 P_{X_1}(-A) P_{X_2}(A|x_1 = -A) \\ & - A^2 P_{X_1}(A) P_{X_2}(-A|x_1 = A). \end{aligned} \tag{2.55}$$

On the other hand,

$$P_{X_2}(A|x_1 = A) = 1 - P_{X_2}(-A|x_1 = A),$$

in which $P_{X_2}(A|x_1 = A)$ is the probability of observing $x_2 = A$ given that $x_1 = A$.

First, consider the condition $\tau < T_b$. In this case, $x_2 = -A$, when $x_1 = A$ only if a transition occurs in the interval $[t, t + \tau]$, and the state changes at this point.

But, because of the assumption of a uniform distribution for the initial transition, the probability that a transition lies in the interval $[t, t + \tau]$ is proportional to the size of the interval, that is, τ/T_b. Hence,

$$P_{X_2}(-A|x_1 = A) = \frac{\tau}{2T_b}, \quad 0 < \tau < T_b,$$

and

$$P_{X_2}(A|x_1 = A) = 1 - \frac{\tau}{2T_b}, \ 0 < \tau < T_b,$$

In a similar way,

$$P_{X_2}(A|x_1 = -A) = \frac{\tau}{2T_b}, \ 0 < \tau < T_b,$$

and

$$P_{X_2}(-A|x_1 = -A) = 1 - \frac{\tau}{2T_b}, \ 0 < \tau < T_b.$$

Substituting into Equation (2.55) and making the simplifications, one obtains

$$R_X(\tau) = 1 - \frac{\tau}{T_b}, \ 0 < \tau < T_b.$$

Therefore, considering that the autocorrelation is an even function, and that the signal pulses are uncorrelated outside the symbol interval T_b, the digital signal $X(t)$ has the following autocorrelation function, for any time difference,

$$R_X(\tau) = A^2 \left[1 - \frac{|\tau|}{T_b}\right] [u(\tau + T_b) - u(\tau - T_b)], \tag{2.56}$$

in which T_b is the pulse duration.

Example: A telegraphic signal $X(t)$, with equiprobable amplitude levels A and $-A$, is illustrated in Figure 2.14. For this signal, the transitions can occur at any time, and they are usually modeled as a Poisson stochastic process.

For the Poisson process, with an arriving rate λ, the probability that the system is in state k, at time τ, is given by the following formula

$$p_k(\tau) = \frac{(\lambda\tau)^k}{k!}e^{-\lambda\tau}. \tag{2.57}$$

Using (2.57), the probability of no transition in the interval is

$$p_0(\tau) = \frac{(\lambda\tau)^0}{0!}e^{-\lambda\tau} = e^{-\lambda\tau}, \tag{2.58}$$

and the probability of, at least, a transition occurring is

$$p_T(\tau) = 1 - e^{-\lambda\tau}. \tag{2.59}$$

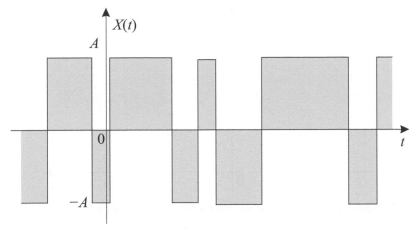

Figure 2.14 Telegraphic random signal.

The autocorrelation of the telegraphic signal can be computed in the following manner. As before,

$$R_X(\tau) = \sum_{X_1} \sum_{X_2} x_1 x_2 P_{X_1 X_2}(x_1, x_2) \tag{2.60}$$

in which x_1 and x_2 represent the signal amplitudes at t and $t + \tau$, and the summations are over the sets $X_1 = X_2 = \{-A, A\}$.

This can again be written as

$$\begin{aligned} R_X(\tau) = {} & A^2 \left[P_{X_1 X_2}(A, A) + P_{X_1 X_2}(-A, -A) \right. \\ & \left. - P_{X_1 X_2}(-A, A) - P_{X_1 X_2}(A, -A) \right]. \end{aligned} \tag{2.61}$$

Substituting the transition probabilities, from the Poisson formula, and simplifying the equation, give

$$R_X(\tau) = A^2 e^{-\lambda |\tau|}. \tag{2.62}$$

The autocorrelation function for the telegraphic signal is shown in Figure 2.15. One can verify that the signal power is $P_X = R_X(0) = A^2$.

2.4.6 The Power Spectral Density

The autocorrelation function makes it possible to define the following Fourier transform pair, known as the Wiener–Khintchin theorem.

$$S_X(w) = \int_{-\infty}^{+\infty} R_X(\tau) e^{-jw\tau} d\tau \tag{2.63}$$

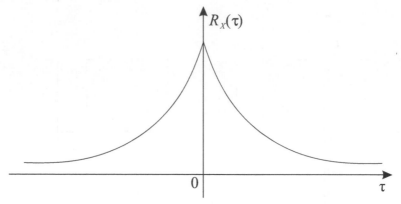

Figure 2.15   Autocorrelation of a telegraphic random signal.

$$R_X(\tau) = \frac{1}{2\pi} \int_{-\infty}^{+\infty} S_X(w)e^{jw\tau}\,dw. \tag{2.64}$$

The function $S_X(w)$ is called the PSD of the random process.

The Wiener–Khintchin theorem relates the autocorrelation function with the PSD, that is, it plays the role of a bridge between the time domain and the frequency domain for random signals. This theorem will be proved in the sequel. Figure 2.16 shows a random signal truncated in an interval T.

The Fourier transform for the signal $x(t)$ given in Figure 2.16 is given by $\mathcal{F}[x_T(t)] = X_T(\omega)$. The time, or deterministic, PSD of $x(t)$ is calculated as follows.

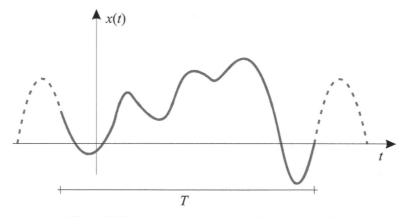

Figure 2.16   Random signal truncated in an interval T.

$$\lim_{T \to \infty} \frac{1}{T} |X_T(\omega)|^2 = \overline{S_X(\omega)}$$

The result obtained is obviously a random quantity, and it is possible to compute its statistical mean to obtain the PSD

$$S_X(\omega) = \mathrm{E}[\overline{S_X(\omega)}] \tag{2.65}$$

Recall that

$$|X_T(\omega)|^2 = X_T(\omega) \cdot X_T^*(\omega) = X_T(\omega) \cdot X_T(-\omega)$$

for a real $X(t)$, in which

$$X_T(\omega) = \int_{-T/2}^{T/2} X_T(t) e^{-j\omega t}\, dt.$$

Substituting the last transform into Equation (2.65), with an adequate variable change, gives

$$\begin{aligned}
S_X(\omega) &= \lim_{T \to \infty} \frac{1}{T} \mathrm{E}[|X_T(\omega)|^2] = \lim_{T \to \infty} \frac{1}{T} \mathrm{E}[X_T(\omega) \cdot X_T(-\omega)] \\
&= \lim_{T \to \infty} \mathrm{E}\left[\int_{-T/2}^{T/2} X_T(\sigma) e^{-j\omega\sigma}\, d\sigma \cdot \int_{-T/2}^{T/2} X_T(\rho) e^{j\omega\rho}\, d\rho \right] \\
&= \lim_{T \to \infty} \frac{1}{T} \int_{-T/2}^{T/2} \int_{-T/2}^{T/2} \mathrm{E}[X_T(\sigma) X_T(\rho)] e^{-j(\sigma-\rho)\omega}\, d\sigma\, d\rho \\
&= \lim_{T \to \infty} \frac{1}{T} \int_{-T/2}^{T/2} \int_{-T/2}^{T/2} R_{X_T}(\sigma - \rho) e^{-j(\sigma-\rho)\omega}\, d\sigma\, d\rho.
\end{aligned}$$

Figure 2.17 shows the integration region. The real plane is covered as $T \to \infty$.

Making the variable changes $\tau = \sigma - \rho$ and $t = \sigma$, the PSD is given by

$$S_X(\omega) = \lim_{T \to \infty} \frac{1}{T} \int_{-T/2}^{T/2} \int_{-T/2-t}^{T/2-t} R_{X_T}(\tau) e^{-j\tau\omega}\, d\tau\, dt.$$

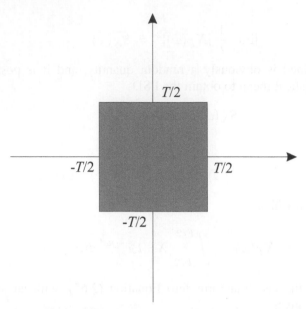

Figure 2.17   Rectangular integration region.

Figure 2.18 shows the new integration region, after the variable changes. In order to compute the integral, with the indicated limits, it is useful to separate the integration region into two parts, the first one for $\tau \geq 0$ and the other for $\tau < 0$.

Therefore, the expression is given by

$$S_X(\omega) = \lim_{T \to \infty} \frac{1}{T} \left[\int_0^T \int_{-T/2}^{T/2-\tau} R_{X_T}(\tau) e^{-j\tau\omega} \, d\tau \, dt \right.$$
$$\left. + \int_{-T}^0 \int_{-T/2-\tau}^{T/2} R_{X_T}(\tau) e^{-j\tau\omega} \, d\tau \, dt \right].$$

Integration of the equation gives

$$S_X(\omega) = \lim_{T \to \infty} \frac{1}{T} \left[\int_0^T (T - \tau) R_{X_T}(\tau) e^{-j\tau\omega} \, d\tau \right.$$
$$\left. + \int_{-T}^0 (T + \tau) R_{X_T}(\tau) e^{-j\tau\omega} \, d\tau \right].$$

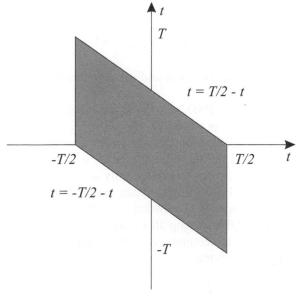

Figure 2.18 Integration region obtained after the variable change.

Combining the integrals, one obtains

$$S_X(\omega) = \lim_{T \to \infty} \frac{1}{T} \left[\int_{-T}^{T} R_{X_T}(\tau) e^{-j\tau\omega} \, d\tau \right.$$
$$\left. - 2 \int_{0}^{T} \tau \cos(\omega\tau) R_{X_T}(\tau) e^{-j\tau\omega} \, d\tau \right].$$

Finally, taking the limit and computing the integral in the second term of the expression, lead to the Wiener–Khintchin theorem

$$S_X(\omega) = \int_{-\infty}^{\infty} R_X(\tau) e^{-j\omega\tau} \, d\tau. \tag{2.66}$$

Therefore, $S_X(\omega) = \mathcal{F}[R_X(\tau)]$, given that τR_X is absolutely integrable, which implies that

$$\int_{-\infty}^{\infty} |\tau R_X(\tau)| \, d\tau < \infty.$$

The function $S_X(\omega)$ represents the PSD, which measures power per unit frequency. The corresponding inverse transform is the autocorrelation function

$R_X(\tau) = \mathcal{F}^{-1}[S_X(\omega)]$, or

$$R_X(\tau) = \frac{1}{2\pi} \int_{-\infty}^{\infty} S_X(\omega)e^{j\omega\tau}\,d\omega. \tag{2.67}$$

Figure 2.19 illustrates the PSD function and the autocorrelation function for a band-limited signal.

Figure 2.20 illustrates the PSD and the corresponding autocorrelation function for white noise. It is noticed that $S_X(\omega) = S_0$, which indicates a uniform distribution for the PSD along the spectrum, and $R_X(\tau) = S_0\delta(\tau)$, which shows white noise as the most uncorrelated, or most random, of all signals. Correlation is non-zero for this signal only at $\tau = 0$.

On the other hand, the PSD for a constant signal is an impulse at the origin, and its autocorrelation is a constant for all τ, which renders the constant as the most predictable among all signals.

Example: The PDF for the telegraphic signal can be computed using the Wiener–Khintchin theorem, giving

$$S_X(\omega) = A^2 \int_{-\infty}^{\infty} R_X(\tau)e^{-j\omega\tau}\,e^{-\lambda|\tau|}d\tau.$$

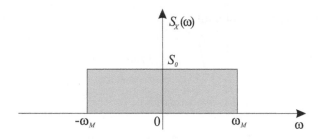

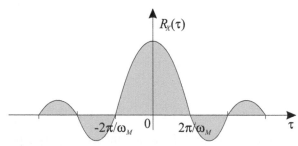

Figure 2.19 Power spectral density function and the autocorrelation function for a band-limited signal.

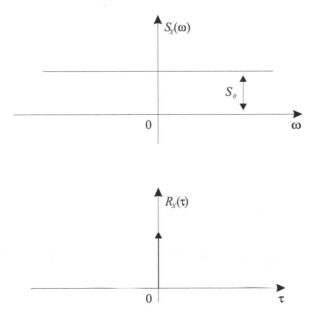

Figure 2.20 Power spectral density function and the autocorrelation function for white noise.

Splitting into two integrals,

$$S_X(\omega) = A^2 \int_0^\infty R_X(\tau)e^{-(j\omega+\lambda)\tau}d\tau + A^2 \int_{-\infty}^0 R_X(\tau)e^{-(j\omega-\lambda)\tau}d\tau.$$

Changing the variables and the order of integration, and computing the integrals,

$$S_X(\omega) = \frac{A^2}{\lambda + j\omega} + \frac{A^2}{\lambda - j\omega}.$$

Computing the common divisor and adding the fractions, one obtains

$$S_X(\omega) = \frac{2\lambda A^2}{\lambda^2 + \omega^2}. \qquad (2.68)$$

The PSD for the telegraphic signal is shown in Figure 2.21.

2.4.7 Propertics of the Power Spectral Density

Some properties of the PSD function are listed in the following.

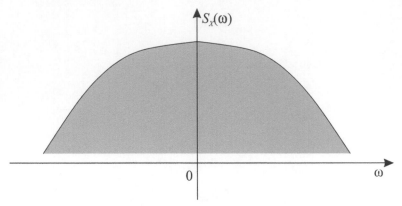

Figure 2.21   Power spectral density of a telegraphic random signal.

1. The area under the curve of the PSD is equal to the total power of the random process, that is,

$$P_X = \frac{1}{2\pi} \int_{-\infty}^{+\infty} S_X(\omega)d\omega. \qquad (2.69)$$

This fact can be verified directly as

$$P_X = R_X(0) = \frac{1}{2\pi} \int_{-\infty}^{\infty} S_X(\omega)e^{j\omega 0}\, d\omega$$

$$= \frac{1}{2\pi} \int_{-\infty}^{\infty} S_X(\omega)\, d\omega = \int_{-\infty}^{\infty} S_X(f)\, df,$$

in which $\omega = 2\pi f$.

2. The area under the autocorrelation function is PSD computed at the origin,

$$S_X(0) = \int_{-\infty}^{\infty} R_X(\tau)\, d\tau. \qquad (2.70)$$

The proof is similar to the previous property.

3. If $R_X(\tau)$ is real and even then

$$S_X(\omega) = \int_{-\infty}^{\infty} R_X(\tau)[\cos \omega\tau - j\sin \omega\tau]\, d\tau,$$

$$= \int_{-\infty}^{\infty} R_X(\tau) \cos \omega\tau\, d\tau, \qquad (2.71)$$

that is, $S_X(\omega)$ is real and even.

4. $S_X(\omega) \geq 0$, since the density reflects a power measure.
5. The following identities hold.

$$\int_{-\infty}^{+\infty} R_X(\tau) R_Y(\tau) d\tau = \frac{1}{2\pi} \int_{-\infty}^{+\infty} S_X(w) S_Y(w) dw \qquad (2.72)$$

$$\int_{-\infty}^{+\infty} R_X^2(\tau) d\tau = \frac{1}{2\pi} \int_{-\infty}^{+\infty} S_X^2(w) dw. \qquad (2.73)$$

Finally, the cross-correlation, or just correlation, between two random processes $X(t)$ and $Y(t)$ is defined as

$$R_{XY}(\tau) = \mathrm{E}[X(t)Y(t+\tau)], \qquad (2.74)$$

The cross-correlation has some important properties that are useful in deriving the quadrature modulation theory:

1. For any two stochastic processes,

$$R_{XY}(-\tau) = \mathrm{E}[X(t)Y(t-\tau)] = \mathrm{E}[X(t+\sigma)Y(t)] = R_{YX}(\tau). \quad (2.75)$$

2. The correlation is bounded by

$$|R_{XY}(\tau)| \leq \sqrt{R_X(0)R_Y(0)}. \qquad (2.76)$$

3. If the two stochastic processes $x(t)$ or $y(t)$ are independent, then

$$R_{XY}(\tau) = R_{YX}(\tau) = \mathrm{E}[X(t)]\mathrm{E}[Y(t)]. \qquad (2.77)$$

The Fourier transform of the correlation leads to the definition of the cross-PSD $S_{XY}(\omega)$.

$$S_{XY}(\omega) = \int_{-\infty}^{+\infty} R_{XY}(\tau) e^{-j\omega\tau} d\tau. \qquad (2.78)$$

Example: By knowing that $\hat{m}(t) = \frac{1}{\pi t} * m(t)$ is the Hilbert transform of $m(t)$ and using properties of the autocorrelation and of the cross-correlation, it can be shown that

$$\mathrm{E}[m(t)^2] = \mathrm{E}[\hat{m}(t)^2]$$

and that

$$\mathrm{E}[m(t)\hat{m}(t)] = 0.$$

The following property is commonly used in communications, to detect a carrier or a pilot signal. If two stationary processes, $X(t)$ and $Y(t)$, are added to form a new process $Z(t) = X(t) + Y(t)$, then the autocorrelation function of the new process is given by

$$R_Z(\tau) = \mathrm{E}[Z(t) \cdot Z(t + \tau)]$$
$$= \mathrm{E}[(x(t) + y(t))(x(t + \tau) + y(t + \tau))],$$

which implies

$$R_Z(\tau) = \mathrm{E}[x(t)x(t + \tau) + y(t)y(t + \tau) + x(t)y(t + \tau) + x(t + \tau)y(t)].$$

By applying properties of the expected value to the above expression, it follows that

$$R_Z(\tau) = R_X(\tau) + R_Y(\tau) + R_{XY}(\tau) + R_{YX}(\tau). \qquad (2.79)$$

If the stochastic processes $X(t)$ and $Y(t)$ can be considered uncorrelated, then $R_{XY}(\tau) = R_{YX}(\tau) = 0$, and $R_Z(\tau)$ is written as

$$R_Z(\tau) = R_X(\tau) + R_Y(\tau). \qquad (2.80)$$

This equation is useful to detect a periodic signal in the presence of noise by correlation.

Example: A radar signal is a periodic waveform that is, for example, transmitted to detect the presence of an object, usually an enemy plane. Of course, the enemy does not want to be detected, and generates an electromagnetic noise signal to counter-attack and hide the plane.

Suppose, for the sake of simplicity, that the transmitted radar signal is a sinusoidal waveform, $X(t) = A\cos(\omega_0 + \phi)$, and the thermal noise $N(t)$ is produced by a device operating at a high temperature.

Then the received signal is

$$Z(t) = A\cos(\omega_0 + \phi) + N(t),$$

and the autocorrelation is given by

$$R_Z(\tau) = \frac{A^2}{2}\cos(\omega_0\tau) + \alpha kTGe^{-\alpha|\tau|},$$

in which α is a parameter that depends on the medium characteristics, the parameter $k = 1.38 \times 10^{-23}$ is the Boltzmann constant, G is the conductance of the device, and T is the absolute temperature, in kelvin.

Figure 2.22 shows that, for large values of τ, the thermal noise fades down, and the autocorrelation of the received signal shows a periodic behavior, of the same period as that of $X(t)$. This permits the detection of the periodic signal.

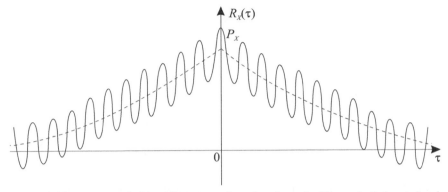

Figure 2.22   Autocorrelation at the output of a noisy channel with a periodic input signal.

The decorrelation time, which is used to estimate how large τ should be to extract the periodic signal, is given by (Lévine, 1973)

$$\tau_0 = \frac{1}{2} \int_{-\infty}^{\infty} |\rho(\tau)| d\tau, \tag{2.81}$$

in which the correlation coefficient is defined as

$$\rho(\tau) = \frac{R(\tau)}{R(0)}, \quad -1 \leq \rho \leq 1.$$

If, in the formula for the noise autocorrelation, the parameter $\alpha \to \infty$, the function becomes an impulse, and the interfering signal is called Additive White Gaussian Noise (AWGN). In this case, for any value of τ, the noise autocorrelation fades down immediately, and the periodic signal can be easily detected.

The resulting detected power can be written as

$$P_Z = R_Z(0) = P_X + P_Y,$$

and for the example

$$P_Z = \frac{A^2}{2} + \alpha k T G.$$

The corresponding PSD is given by

$$S_Z(\omega) = S_X(\omega) + S_Y(\omega), \tag{2.82}$$

which results in

$$S_Z(\omega) = \frac{\pi A^2}{2} [\delta(\omega + \omega_0) + \delta(\omega - \omega_0)] + \frac{2\alpha^2 k T G}{\alpha^2 + \omega^2},$$

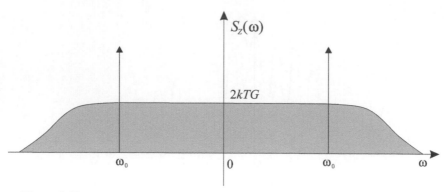

Figure 2.23   Spectrum at the output of a noisy channel with a periodic input signal.

that is shown in Figure 2.23.

2.5 Linear Systems

Linear systems can be analyzed using the theory of stochastic processes. This provides more general and more interesting solutions than those resulting from classical analysis. This section deals with the response of linear systems to a random input $X(t)$.

For a linear system, as illustrated in Figure 2.24, the Fourier transform of its impulse response $h(t)$ is given by

$$H(\omega) = \int_{-\infty}^{\infty} h(t)e^{-j\omega t} \, dt. \qquad (2.83)$$

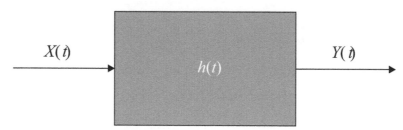

Figure 2.24   Linear system fed with a random input signal.

The linear system response $Y(t)$ is obtained by means of the convolution of the input signal with the impulse response as follows.

$$Y(t) = X(t) * h(t) \Rightarrow Y(t) = \int_{-\infty}^{\infty} X(t - \alpha) h(\alpha) \, d\alpha$$

$$= \int_{-\infty}^{\infty} X(\alpha) h(t - \alpha) \, d\alpha.$$

2.5.1 Expected Value of the Output Signal

The mean value of the random signal at the output of a linear system is calculated as follows.

$$E[Y(t)] = E\left[\int_{-\infty}^{\infty} X(t - \alpha) h(\alpha) \, d\alpha\right] = \int_{-\infty}^{\infty} E[X(t - \alpha)] h(\alpha) \, d\alpha$$

Considering the random signal $X(t)$ to be narrow-sense stationary, it follows that $E[X(t - \alpha)] = E[X(t)] = m_X$, and thus

$$E[Y(t)] = m_X \int_{-\infty}^{\infty} h(\alpha) \, d\alpha = m_X H(0),$$

in which $H(0) = \int_{-\infty}^{\infty} h(\alpha) \, d\alpha$ follows from (2.83) computed at $\omega = 0$. Therefore, the mean value of the output signal depends only on the mean value of the input signal and on the value assumed by the transfer function at $\omega = 0$.

2.5.2 The Response of Linear Systems to Random Signals

The computation of the autocorrelation of the output signal, given the auto-correlation of the input signal to a linear system, can be performed as follows.

The relationship between the input and the output of a linear system was shown earlier to be given by

$$Y(t) = \int_{-\infty}^{\infty} X(\rho) h(t - \rho) \, d\rho = \int_{-\infty}^{\infty} X(t - \rho) h(\rho) \, d\rho = X(t) * h(t).$$

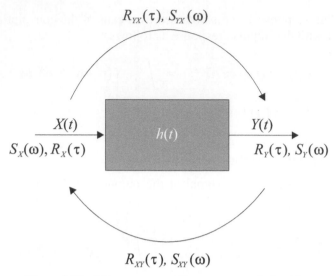

$$R_{YX}(\tau),\, S_{YX}(\omega)$$

$X(t)$
$S_x(\omega),\, R_x(\tau)$

$h(t)$

$Y(t)$
$R_Y(\tau),\, S_Y(\omega)$

$$R_{XY}(\tau),\, S_{XY}(\omega)$$

Figure 2.25   Linear system and input–output relationships.

The output autocorrelation function can be calculated directly from its definition as

$$
\begin{aligned}
R_Y(\tau) &= \mathrm{E}[Y(t)Y(t+\tau)] \\
&= \mathrm{E}\left[\int_{-\infty}^{\infty} X(t-\rho)\,h(\rho\,d\rho \cdot \int_{-\infty}^{\infty} X(t+\tau-\sigma)\,h(\sigma)\,d\sigma\right] \\
&= \int_{-\infty}^{\infty}\int_{-\infty}^{\infty} \mathrm{E}[X(t-\rho)X(t+\tau-\sigma)] \cdot h(\rho) \cdot h(\sigma)\,d\rho\,d\sigma \\
&= \int_{=\infty}^{\infty}\int_{-\infty}^{\infty} R_{XX}(\tau+\rho-\sigma)\,h(\rho)\,h(\sigma)\,d\rho\,d\sigma.
\end{aligned}
$$

Example: Suppose that white noise with autocorrelation $R_X(\tau) = \delta(\tau)$ is the input signal to a linear system. The corresponding autocorrelation function of the output signal is given by

$$
\begin{aligned}
R_Y(\tau) &= \int_{-\infty}^{\infty}\int_{-\infty}^{\infty} \delta(\tau+\rho-\sigma)\,h(\rho)\,h(\sigma)\,d\rho\,d\sigma \\
&= \int_{-\infty}^{\infty} h(\sigma-\tau) \cdot h(\sigma)\,d\sigma \\
&= h(-\tau) * h(\tau).
\end{aligned}
$$

The Fourier transform of $R_Y(\tau)$ leads to the following result

$$R_Y(\tau) = h(-t) * h(t) \iff S_Y(\omega) = H(-\omega) \cdot H(\omega),$$

and for $h(\tau)$ a real function of τ, it follows that $H(-\omega) = H^*(\omega)$, and consequently

$$S_Y(\omega) = H(-\omega) \cdot H(\omega) = H^*(\omega) \cdot H(\omega) = |H(\omega)|^2.$$

Summarizing, the output PSD is $S_Y(\omega) = |H(\omega)|^2$ when white noise is the input to a linear system.

In general, the output spectrum can be computed by applying the Wiener–Khintchin theorem $S_Y(\omega) = \mathcal{F}[R_Y(\tau)]$,

$$S_Y(\omega) = \int_{-\infty}^{\infty} \int_{-\infty}^{\infty} \int_{-\infty}^{\infty} R_X(\tau + \rho - \sigma) \, h(\rho) \, h(\sigma) \cdot e^{-j\omega\tau} \, d\rho \, d\sigma \, d\tau.$$

Integrating on the variable τ, it follows that

$$S_Y(\omega) = \int_{-\infty}^{\infty} \int_{-\infty}^{\infty} S_X(\omega) e^{j\omega(\rho-\sigma)} \, h(\rho) \, h(\sigma) \, d\rho \, d\sigma.$$

Finally, removing $S_X(\omega)$ from the double integral and then separating the two variables in this double integral, it follows that

$$S_Y(\omega) = S_X(\omega) \int_{-\infty}^{\infty} h(\rho) e^{j\omega\rho} \, d\rho \int_{-\infty}^{\infty} h(\sigma) e^{-j\omega\sigma} \, d\sigma$$
$$= S_X(\omega) \cdot H(-\omega) \cdot H(\omega).$$

Therefore, $S_Y(\omega) = S_X(\omega) \cdot |H(\omega)|^2$ will result whenever the system impulse response is a real function.

Example: Consider again white noise with autocorrelation function $R_X(\tau) = \delta(\tau)$ applied to a linear system. The white noise spectrum is calculated as follows

$$S_X(\omega) = \int_{-\infty}^{\infty} R_X(\tau) e^{-j\omega\tau} \, d\tau = \int_{-\infty}^{\infty} \delta(\tau) e^{-j\omega\tau} \, d\tau = 1,$$

from which it follows that

$$S_Y(\omega) - |II(\omega)|^2,$$

similar to the previous example.

Example: The linear system shown in Figure 2.26 is a differentiator, used in control systems or demodulator/detector for frequency-modulated signals.

The output PSD for this circuit (or its frequency response) is equal to

$$S_Y(\omega) = |j\omega|^2 \cdot S_X(\omega) = \omega^2 S_X(\omega).$$

It is thus noticed that, for frequency-modulated signals, the noise at the detector output follows a square law, that is, the output PSD grows with the square of the frequency. In this manner, in a frequency division multiplexing of frequency-modulated channels, the noise will affect more intensely those channels occupying the higher frequency region of the spectrum.

Figure 2.27 shows, as an illustration of what has been discussed so far about square noise, the spectrum of a low-pass flat noise (obtained by passing white noise through an ideal low-pass filter). This filtered white noise is applied to the differentiator circuit of the example, which in turn produces at the output the square law noise shown in Figure 2.28.

Observation: Pre-emphasis circuits are used in FM modulators to compensate for the effect of square noise effect.

Other relationships among different correlation functions can be derived, as illustrated in Figure 2.25. The correlation measures between input and

$X(t)$

$R_x(\tau)$

$d(\cdot)/dt$

$X(t)$

$R_y(\tau)$

Figure 2.26   A differentiator circuit.

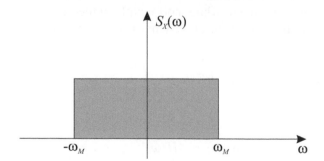

$S_x(\omega)$

$-\omega_M$

ω_M

ω

Figure 2.27   Spectrum of the low-pass noise.

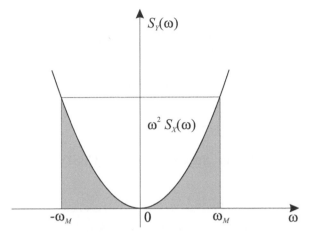

Figure 2.28 Spectrum of the quadratic noise.

output (input–output cross-correlation) and correlation between output and input can also be calculated from the input signal autocorrelation.

The correlation between the input and the output can be calculated with the formula

$$R_{XY}(\tau) = \mathrm{E}[X(t)Y(t+\tau)],$$

and in an analogous manner, the correlation between the output and the input can be calculated as

$$R_{YX}(\tau) = \mathrm{E}[Y(t)X(t+\tau)].$$

For a linear system, the correlation between output and input is given by

$$R_{YX}(\tau) = \mathrm{E}\left[\int_{-\infty}^{\infty} X(t-\rho)\,h(\rho)\,d\rho \cdot X(t+\tau)\right].$$

Exchanging the order of the expected value and integral computations, due to their linearity, it follows that

$$R_{YX}(\tau) = \int_{-\infty}^{\infty} \mathrm{E}[X(t-\rho)X(t+\tau)]\,h(\rho)\,d\rho = \int_{-\infty}^{\infty} R_X(\tau+\rho)\,h(\rho)\,d\rho.$$

In a similar manner, the correlation between the input and the output is calculated as

$$R_{XY}(\tau) = \mathrm{E}\left[X(t)\cdot\int_{-\infty}^{\infty} X(t+\tau-\rho)\,h(\rho)\,d\rho\right]$$
$$= \int_{-\infty}^{\infty} \mathrm{E}[X(t)X(t+\tau-\rho)]\,h(\rho)\,d\rho,$$

and finally,

$$R_{XY}(\tau) = \int_{-\infty}^{\infty} R_X(\tau - \rho)\, h(\rho)\, d\rho.$$

Therefore

$$R_{XY}(\tau) = R_X(\tau) * h(\tau)$$

and

$$R_{YX}(\tau) = R_X(\tau) * h(-\tau).$$

The resulting Cross-Power Spectral Density (CPSD) between input–output is given by

$$S_{XY}(\tau) = S_X(\omega) \cdot H(\omega),$$

and between output–input is

$$S_{YX}(\tau) = S_X(\omega) \cdot H^*(\omega).$$

By assuming $S_Y(\omega) = |H(\omega)|^2 S_X(\omega)$, the following relationships are immediate.

$$\begin{aligned} S_Y(\omega) &= H^*(\omega) \cdot S_{XY}(\omega) \\ &= H(\omega) \cdot S_{YX}(\omega). \end{aligned}$$

It is usually easier to deal with PSDs rather than autocorrelations, and the determination of $S_Y(\omega)$ from $S_X(\omega)$ is a rather direct operation, for linear systems. The CPSDs are calculated afterward, and the correlations are then obtained by inverse Fourier transformation. This is indicated in the following.

$$\begin{array}{ccccc}
R_X(\tau) & \longleftrightarrow & S_X(\omega) & & \\
& & \downarrow & & \\
R_Y(\tau) & \longleftrightarrow & S_Y(\omega) & \longrightarrow \quad S_{XY}(\omega) \longleftrightarrow R_{XY}(\tau) \\
& & \downarrow & & \\
& & S_{YX}(\omega) & & \\
& & \updownarrow & & \\
& & R_{YX}(\tau) & &
\end{array}$$

2.5.3 Phase Information

The autocorrelation is a special measure of average behavior of a signal. Consequently, it is not always possible to recover a signal from its autocorrelation.

Since the PSD is a function of the autocorrelation, it also follows that signal recovery from its PSD is not always possible because phase information about the signal has been lost in the averaging operation involved.

On the other hand, the CPSDs, relating input–output and output–input, preserve signal phase information and can be used to recover the phase function explicitly.

The transfer function of a linear system can be written as

$$H(\omega) = |H(\omega)|e^{j\theta(\omega)},$$

in which the modulus $|H(\omega)|$ and the phase $\theta(\omega)$ are clearly separated. The complex conjugate of the transfer function is

$$H^*(\omega) = |H(\omega)|e^{-j\theta(\omega)}.$$

Since

$$S_Y(\omega) = H^*(\omega)S_{XY}(\omega) = H(\omega) \cdot S_{YX}(\omega),$$

it follows for a real $h(t)$ that

$$|H(\omega)|e^{-j\theta\omega} \cdot S_{XY}(\omega) = |H(\omega)| \cdot e^{j\theta\omega} \cdot S_{YX}(\omega)$$

and finally,

$$e^{2j\theta(\omega)} = \frac{S_{XY}(\omega)}{S_{YX}(\omega)}.$$

The function $\theta(\omega)$ can then be extracted thus giving

$$\theta(\omega) = \frac{1}{2j} \ln \frac{S_{XY}(\omega)}{S_{YX}(\omega)},$$

which is the desired signal phase information.

Thus, the previous diagram can be completed with the information on how to compute the phase function of the signal, as follows.

$$
\begin{array}{ccccc}
R_X(\tau) & \longleftrightarrow & S_X(\omega) & & \\
 & & \downarrow & & \\
R_Y(\tau) & \longleftrightarrow & S_Y(\omega) & \longrightarrow \ S_{XY}(\omega) & \longleftrightarrow \ R_{XY}(\tau) \\
 & & \downarrow & \downarrow & \\
 & & S_{YX}(\omega) & \longrightarrow \ \theta(\omega) & \\
 & & \updownarrow & & \\
 & & R_{YX}(\tau) & &
\end{array}
$$

Example: The Hilbert transform provides an example of how to apply the preceding theory. A representation of the filter impulse response for

$X(t)$ $1/\pi t$ $Y(t)$

Figure 2.29   Hilbert filter with a random input signal.

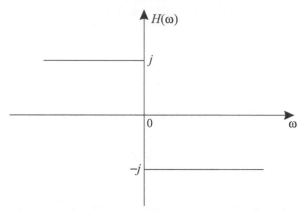

Figure 2.30   Transfer function of the Hilbert transform.

the Hilbert transform is shown in Figure 2.29. The frequency domain representation of the Hilbert transform is shown in Figures 2.30 and 2.31.

Since for the Hilbert transform $H(\omega) = j\left[u(-\omega) - u(\omega)\right]$, it follows that $|H(\omega)|^2 = 1$, and from the important result

$$S_Y(\omega) = |H(\omega)|^2 \cdot S_X(\omega),$$

it follows that

$$S_Y(\omega) = S_X(\omega).$$

Therefore, the PSD of a signal at the output of a Hilbert filter is exactly the same as the input. The fact that $S_Y(\omega) = S_X(\omega)$ is expected, since the Hilbert transform operates only on the signal phase and the PSD does not contain phase information.

2.6 Analysis of a Digital Signal

Most modulation techniques use digital signals as inputs, and this is the only reason they are called digital modulation schemes. Then, digital signal analysis is important to obtain the spectra of digitally modulated carriers.

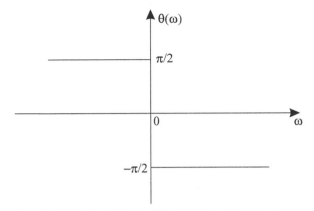

Figure 2.31 The phase function of the Hilbert transform, in the frequency domain.

This section presents a mathematical formulation for the digital signal, which includes the computation of its autocorrelation and power density spectrum.

The digital signal can be produced by the digitization of an audio or a video signal, or can be directly generated by a digital equipment, or by a computer connected to a network. A possible mathematical expression of the digital signal is as follows,

$$m(t) = \sum_{k=-\infty}^{\infty} m_k p(t - kT_b), \qquad (2.84)$$

in which m_k represents the k-th randomly generated symbol from a discrete alphabet M, $p(t)$ is the pulse function that shapes the transmitted signal, and T_b is the bit interval.

2.6.1 Autocorrelation of a Digital Signal

As previously discussed, the autocorrelation function for signal $m(t)$, which can be non-stationary, is computed by the formula

$$R_M(\tau, t) = \mathrm{E}[m(t)m(t + \tau)]. \qquad (2.85)$$

Substituting the expression for $m(t)$ into Formula (2.85) gives

$$R_M(\tau, t) = \mathrm{E}\left[\sum_{k=-\infty}^{\infty} \sum_{i=-\infty}^{\infty} m_k p(t - kT_b) m_j p(t + \tau - iT_b) \right]. \qquad (2.86)$$

Because of the linearity property, the expected value operator applies directly to the random signals,

$$R_M(\tau, t) = \sum_{k=-\infty}^{\infty} \sum_{i=-\infty}^{\infty} \mathrm{E}\left[m_k m_j\right] p(t - kT_b) p(t + \tau - iT_b). \qquad (2.87)$$

Equation (2.87) is averaged in time, to eliminate the time dependency, producing

$$R_M(\tau) = \frac{1}{T_b} \int_0^{T_b} R_M(\tau, t) dt, \qquad (2.88)$$

or, equivalently

$$R_M(\tau) = \frac{1}{T_b} \int_0^{T_b} \sum_{k=-\infty}^{\infty} \sum_{i=-\infty}^{\infty} \mathrm{E}\left[m_k m_j\right] p(t - kT_b) p(t + \tau - iT_b) dt. \qquad (2.89)$$

The integral and summation operations can be changed, which gives

$$R_M(\tau) = \frac{1}{T_b} \sum_{k=-\infty}^{\infty} \sum_{i=-\infty}^{\infty} \mathrm{E}\left[m_k m_j\right] \int_0^{T_b} p(t - kT_b) p(t + \tau - iT_b) dt. \qquad (2.90)$$

In the last equation, it is possible to define the discrete autocorrelation, to simplify the expression, as

$$R(k - i) = \mathrm{E}\left[m_k m_j\right]. \qquad (2.91)$$

The signal autocorrelation can be written as

$$R_M(\tau) = \frac{1}{T_b} \sum_{k=-\infty}^{\infty} \sum_{i=-\infty}^{\infty} R(k - i) \int_0^{T_b} p(t - kT_b) p(t + \tau - iT_b) dt, \qquad (2.92)$$

which is the general formula for the autocorrelation of a digital signal.

For the particular case of a signal, shaped as a rectangular pulse defined in the interval $0 \le t \le T_b$, with independent and equiprobable symbols from the set $\mathbb{M} = \{-A, A\}$, the autocorrelation function is given by

$$R_M(\tau) = A^2 [1 - \frac{|\tau|}{T_b}][u(\tau + T_b) - u(\tau - T_b)], \qquad (2.93)$$

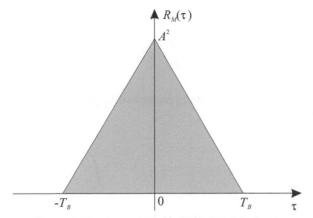

Figure 2.32   Autocorrelation for the digital signal.

in which T_b is the bit interval and A represents the pulse amplitude.

Figure 2.32 shows that the autocorrelation function has a triangular shape. Its maximum, the signal power, occurs at the origin, and is equal to A^2. The autocorrelation decreases linearly with the time interval, and reaches zero at time T_b.

2.6.2 Power Spectral Density for the Digital Signal

The Fourier transform of the autocorrelation function, Equation (2.92), is computed to obtain the PSD for the digital signal.

$$S_M(\omega) = \int_{-\infty}^{\infty} R_M(\tau)e^{-j\omega\tau}d\tau, \tag{2.94}$$

Substituting the expression for the autocorrelation,

$$S_M(\omega) = \frac{1}{T_b} \sum_{k=-\infty}^{\infty} \sum_{i=-\infty}^{\infty} R(k-i)$$

$$\cdot \int_{-\infty}^{\infty} \int_{0}^{T_b} p(t-kT_b)p(t+\tau-iT_b)e^{-j\omega\tau}dtd\tau. \tag{2.95}$$

The order of integration is changed, to compute the Fourier integral of the shifted pulse. This can be written as,

$$S_M(\omega) = \frac{1}{T_b} \sum_{k=-\infty}^{\infty} \sum_{i=-\infty}^{\infty} R(k-i) \int_0^{T_b} p(t-kT_b)P(\omega)e^{-j\omega(kT_b-t)}\,dt.$$

(2.96)

Because the term $P(\omega)e^{-j\omega kT_b}$ is independent of time, it can be taken out of the integral, that is,

$$S_M(\omega) = \frac{1}{T_b} \sum_{k=-\infty}^{\infty} \sum_{i=-\infty}^{\infty} R(k-i)P(\omega)e^{-j\omega kT_b} \int_0^{T_b} p(t-kT_b)e^{j\omega t}\,dt.$$

(2.97)

The integral in (2.97) is then evaluated, to give

$$S_M(\omega) = \frac{1}{T_b} \sum_{k=-\infty}^{\infty} \sum_{i=-\infty}^{\infty} R(k-i)P(\omega)P(-\omega)e^{-j\omega(k-j)T_b}.$$

(2.98)

The shape of the spectrum for the random digital signal depends on the pulse shape, defined by $P(\omega)$, and also on the manner the symbols relate to each other, specified by the discrete autocorrelation function $R(k-i)$.

Therefore, the signal design, necessary to produce an adequate modulation scheme, involves pulse shaping as well as the control of the correlation between the transmitted symbols. It can be obtained by signal processing.

For a real pulse, $P(-\omega) = P^*(\omega)$, and the PSD can be written as

$$S_M(\omega) = \frac{|P(\omega)|^2}{T_b} \sum_{k=-\infty}^{\infty} \sum_{i=-\infty}^{\infty} R(k-i)e^{-j\omega(k-j)T_b},$$

(2.99)

which can be simplified to

$$S_M(\omega) = \frac{|P(\omega)|^2}{T_b} S(\omega),$$

(2.100)

if one puts $l = k - i$, the summations can be simplified and the PSD for the discrete sequence of symbols is given by

$$S(\omega) = \sum_{l=-\infty}^{\infty} R(l)e^{-j\omega l T_b}.$$

(2.101)

For the example of Equation (2.93), the corresponding spectrum is

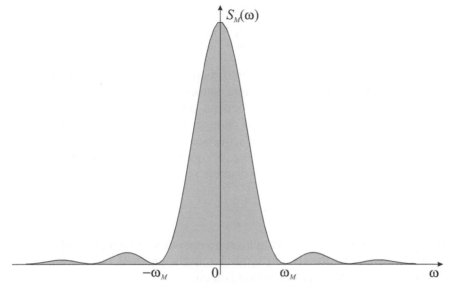

Figure 2.33 Power spectral density for the random digital signal.

$$S_M(\omega) = A^2 T_b \frac{\sin^2 (\omega T_b/2)}{(\omega T_b/2)^2}, \tag{2.102}$$

which is the squared sampling function, and shows that the random digital signal has a continuous spectrum that occupies a large portion of the spectrum. The function is sketched in Figure 2.33. The first null is a usual measure of the bandwidth, and is given by $\omega_M = 2\pi/T_b$.

The previous result has been obtained for a digital signal with equiprobable levels. If level A has probability p, and level $-A$ has probability $1 - p$, then the PSD can be computed, and gives (Gagliardi, 1988)

$$S_M(\omega) = [4p(1-p)] A^2 T_b \frac{\sin^2 (\omega T_b/2)}{(\omega T_b/2)^2} + 2\pi A^2 (2p - 1)^2 \delta(\omega). \tag{2.103}$$

This represents a similar spectrum, with the amplitude adjusted by the factor $[4p(1-p)]$, and with an additional impulse at the origin, indicating a possible DC level, depending on the symbol probability.

It is also possible to shape the signal spectrum by controlling the symbol sequence statistics. First, substitute Equation (2.101) into (2.100).

$$S_M(\omega) = \frac{|P(\omega)|^2}{T_b} \sum_{l=-\infty}^{\infty} R(l)e^{-j\omega l T_b}. \qquad (2.104)$$

The summation in Equation (2.104), which shapes the pulse spectrum, is the discrete Fourier transform of the discrete autocorrelation $R(l)$. Therefore, as the properties of the binary signal depend on the characteristics of the information sequence, is it possible to change the signal spectrum by the introduction of correlation between the sequence bits.

Example: The input bit sequence, m_k, is equally likely and independent, as assumed previously. Suppose a new discrete sequence is produced as

$$b_k = \frac{m_k + m_{k-1}}{2}. \qquad (2.105)$$

Equation (2.105) represents a moving average operation, in which symbol b_k is obtained by averaging the current symbol with the previous one.

The discrete autocorrelation function is then

$$R(l) = \delta(l) + \frac{\delta(l+1) + \delta(l-1)}{2}, \qquad (2.106)$$

in which $\delta(\cdot)$ is the Kronecker delta function, after the German mathematician Leopold Kronecker (1823–1891), defined as $\delta(l) = 1$, for $l = 0$ and zero otherwise.

And the discrete Fourier transform is

$$\begin{aligned} S(\omega) &= \sum_{l=-\infty}^{\infty} R(l)e^{-j\omega l T_b} = 1 + \frac{e^{-j\omega T_b} + e^{j\omega T_b}}{2} \\ &= 1 + \cos\omega T_b \\ &= 2\cos^2 \frac{\omega T_b}{2}. \end{aligned}$$

Finally, the signal spectrum is given by

$$S_M(\omega) = \frac{2|P(\omega)|^2}{T_b} \cos^2 \frac{\omega T_b}{2}. \qquad (2.107)$$

It is possible to note that the pulse spectrum has been low-pass filtered by the sequence spectrum, as a result of the introduction of correlation between the symbols, and also there are no discrete lines in the resulting spectrum.

2.6.3 The Digital Signal Bandwidth

The signal bandwidth can be defined in several ways. In industry, the most usual definition is the half-power bandwidth (ω_{3dB}). This bandwidth is computed by dividing the maximum of the by two, and finding the frequency for which this particular value occurs.

The root-mean-square (RMS) bandwidth is computed using the frequency deviation around the carrier, if the signal is modulated, or around the origin, for a baseband signal. The frequency deviation, or RMS bandwidth, for a baseband signal is given by

$$\omega_{RMS} = \sqrt{\frac{\int_{-\infty}^{\infty} \omega^2 S_M(\omega)d\omega}{\int_{-\infty}^{\infty} S_M(\omega)d\omega}}, \tag{2.108}$$

and the RMS bandwidth is $2\omega_{RMS}$.

The previous formula is equivalent to

$$\omega_{RMS} = \sqrt{\frac{-R_M''(0)}{R_M(0)}}. \tag{2.109}$$

The RMS bandwidth for a modulated signal, around the carrier frequency ω_c, is given by

$$\omega_{RMS} = \sqrt{\frac{\int_{-\infty}^{\infty} (\omega - \omega_c)^2 S_M(\omega - \omega_c)d\omega}{\int_{-\infty}^{\infty} S_M(\omega - \omega_c)d\omega}}. \tag{2.110}$$

The white noise bandwidth can be computed by equating the maximum of the signal power spectrum density to the noise power spectrum density $S_N = \max S_M(\omega)$. After that, the power for both signal and noise are equated and ω_N is obtained. The noise power is $P_N = 2\omega_N S_N$ and the signal power, P_M, is given by the formula

$$P_M = R_M(0) = \frac{1}{2\pi} \int_{-\infty}^{\infty} S_M(\omega)d\omega. \tag{2.111}$$

There are signals that exhibit a finite bandwidth, when the spectrum vanishes after a certain frequency. They exhibit a real bandwidth. Finally, the percent bandwidth is computed by finding the frequency range that includes a known percentage, for instance, 90%, of the signal power.

2.7 Non-Linear Systems

There is no general formula for the response of non-linear circuits to random signals. Every problem has to be dealt with in a specific manner, and some constraints are usually established in order to compute the autocorrelation and the PSD of the output signal.

2.7.1 The Two-Level Quantizer

The objective of this section is to present a procedure for obtaining the autocorrelation and spectrum for a two-level quantizer. The model assumes a Gaussian signal applied to the input of the quantizer.

Some models explain the role of the quantization noise in the quantizing process (Gersho, 1978). The assumption of a large number of quantization levels is always a requisite for most models presented (Alencar, 1993). The non-overload restriction is usually considered in the analysis, as well as the independence between the applied signal and the output noise (Gray, 1990). In this deduction, the noise is not considered independent of the applied signal (Alencar and Scharcanski, 1995).

In the quantization process, a mapping is established between the continuous amplitude domain and the discrete amplitude domain (Sripad and Snyder, 1977). The objective of this section is to present a procedure for obtaining the power spectrum density of the quantization noise, for a two-level quantizer. The model assumes a stochastic signal applied to the input of the quantizer. The input signal is considered Gaussian and stationary in the wide sense.

The lowest possible number of levels is assumed, in order to obtain an upper limit on the effect of the quantization noise. This assumption implies that, in this case, the noise cannot be considered independent of the applied signal, and an analytical expression is found for the quantization noise autocorrelation and spectrum. The analytical approach can be applied to the analysis of robust modulation schemes, such as the delta modulator (DM) or the $\Sigma - \Delta$ modulator (Gray, 1987, 1989).

2.7.2 Quantization Noise Spectrum for a Two-level Quantizer

In this section, the analytical solution to the autocorrelation and power spectrum density of the quantization noise, for a two-level quantizer, is presented. The model used assumes a Gaussian stochastic signal applied to the input of the quantizer. The input signal is considered zero mean and stationary in the wide sense.

The spectrum of the quantization noise is not considered independent of the spectrum of the applied signal, as most models assume, and the signal is permitted to overload the quantizer.

The output of the two-level quantizer is related to the input by the equation

$$y(t) = 2du(x(t)) - d \tag{2.112}$$

in which $u(\cdot)$ is the unit step function, $x(t)$ is the input signal, and d represents the quantization step size.

The autocorrelation function for $y(t)$ is defined by

$$R_Y(\tau) = \mathrm{E}[y(t)y(t+\tau)] \tag{2.113}$$

which can be evaluated by substituting the expression for the joint Gaussian probability function, and by a change of coordinates. This leads to

$$R_Y(\tau) = \frac{2d^2}{\pi \sigma_X^2 \sqrt{1-\rho^2}} \int_0^{\pi/2} \int_0^\infty r e^{-\frac{r^2}{2\pi(1-\rho^2)}[1-\rho \sin 2\theta]} dr d\theta - d^2 \tag{2.114}$$

Integrating the expression with respect the variable r gives

$$R_Y(\tau) = \frac{2d^2(1-\rho^2)}{\pi \sqrt{1-\rho^2}} \int_0^{\pi/2} \frac{1}{1-\rho \sin 2\theta} d\theta - d^2 \tag{2.115}$$

Another integration is performed, with respect to θ, until one finds the final result for the autocorrelation function at the output of a two-level quantizer

$$R_Y(\tau) = \frac{2d^2}{\pi} \arcsin\left(\frac{R_X(\tau)}{\sigma_X^2}\right). \tag{2.116}$$

The solution to this integral is the well-known theorem by Middleton and Vleck (van Vleck and Middleton, 1966), which can be put in a series form using the hypergeometric function expansion (Gradshteyn and Ryzhik, 1990).

$$R_Y(\tau) = \frac{2d^2}{\pi} \sum_{k=0}^\infty \frac{(2k)!}{2^{2k}(k!)^2(2k+1)} \left[\frac{R_X(\tau)}{\sigma_X^2}\right]^{2k+1} \tag{2.117}$$

From this expression, it is possible to extract the terms that are related to the signal and noise, as follows

$$R_Y(\tau) = \frac{2d^2}{\pi \sigma_X^2} R_X(\tau) + R_N(\tau). \tag{2.118}$$

in which the autocorrelation of the quantization noise is given by

$$R_N(\tau) = \frac{2d^2}{\pi} \sum_{k=1}^{\infty} \frac{(2k)!}{2^{2k}(k!)^2(2k+1)} \left[\frac{R_X(\tau)}{\sigma_X^2}\right]^{2k+1}. \qquad (2.119)$$

From Equation (2.118) one realizes that the quantization noise power comes directly from the input signal power. In this case, the model that assumes the output signal as the sum of the input plus independent noise is not feasible any more. Signal and noise are strongly correlated in the two-level quantizer.

The use of the Wiener–Khintchin theorem leads directly to the calculation of the power spectrum density for the quantization noise

$$S_N(\omega) = 4d^2 \sum_{k=1}^{\infty} \frac{(2k)!}{(2\pi)^{2k+1} 2^{2k}(k!)^2(2k+1)\sigma_X^{4k+2}} \\ \cdot \underbrace{S_X(\omega) * \cdots * S_X(\omega)}_{2k+1 \; times}. \qquad (2.120)$$

The autocorrelation functions for the signal and quantization noise, for an input signal with a band-limited flat spectrum, are depicted in Figure 2.34.

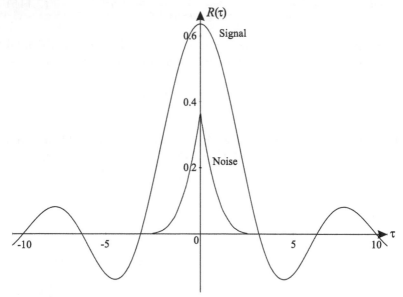

Figure 2.34   Autocorrelation functions for signal and quantization noise.

The noise power at the output of the quantizer is obtained from Equation (2.119), for $\tau = 0$

$$P_N = R_Y(0) - \frac{2d^2}{\pi\sigma_X^2}R_X(0). \tag{2.121}$$

Actually, the net noise power that interferes with the signal is given by

$$P_N = \frac{2d^2}{\pi}\sum_{k=1}^{\infty}\frac{(2k)!}{(2\pi)^{2k+1}2^{2k}(k!)^2(2k+1)\sigma_X^{4k+4}}$$

$$\cdot\int_{-\omega_M}^{\omega_M}\underbrace{S_X(\omega)*\cdots*S_X(\omega)}_{2k+1\ times}\,d\omega. \tag{2.122}$$

in which the integration is limited by the cutoff frequency of the input signal.

From the preceding results and Equation (2.117), one concludes that the output signal retains only 64% of the total input power, and the quantization noise contributes with approximately 36% of the power of the output signal. In fact, the net noise power inside the signal bandwidth is smaller than this, as expressed by Equation (2.122).

Therefore, the upper limit for the normalized quantization noise power is exactly given by $1 - \frac{2}{\pi}$, from Equation (2.118). For a Gaussian signal, this implies, for the worst case, an SQNR of 2.44 dB. The analytical approach used could be applied to the analysis of the DM or the $\Sigma - \Delta$ modulator (Gray, 1989).

2.7.3 Response of a Squarer Circuit

This section analysis the squarer circuit, a device that is used for power measurement, to compute the autocorrelation function and the PSD at the output. The squarer is modeled by (Alencar, 2009b).

$$Y(t) = \beta X^2(t), \tag{2.123}$$

in which β is a parameter of the non-linear circuit.

The autocorrelation of the output signal is

$$R_Y(\tau) = E[Y(t)Y(t+\tau)] = \beta^2 E[X^2(t)X^2(t+\tau)]. \tag{2.124}$$

For a Gaussian zero-mean input signal, it is possible to use Price's theorem (Price, 1958), to obtain the autocorrelation

$$E[X(t)^2X(t+\tau)^2] - 4\int_0^{C_X}E[X(t)X(t+r)]dC(X(t),X(t+r))$$

$$+ E[X(t)]\cdot E[X(t+\tau)], \tag{2.125}$$

in which $C_X = C[X(t), X(t+\tau)]$ is the covariance between the variables $X(t)$ and $X(t+\tau)$.

Computing the integral, one obtains

$$R_Y(\tau) = \beta^2 P_X^2 + 2\beta^2 R_X^2(\tau). \qquad (2.126)$$

Application of the Wiener–Khintchin theorem gives,

$$S_Y(\omega) = \beta^2 P_X^2 \delta(\omega) + \frac{2\beta^2}{2\pi} S_X(\omega) * S_X(\omega). \qquad (2.127)$$

The spectrum shows an impulse at the origin, that corresponds to a DC level introduced by the squarer, in addition to a convolution of the input signal spectra. The convolution spreads the spectrum and, as a consequence, increases the signal bandwidth.

2.7.4 Response of a Non-Linear Amplifier

A non-linear amplifier can be modeled by the following equation, that is a good approximation to the field effect transistor (FET), and also models a radio or television power amplifier that relies on a vacuum tube. Such circuits are also present in several communication, control, and signal processing systems.

The non-linear amplifier can be modeled as a polynomial, considering the first- and second-order terms. Because the device has a random signal at the input, the autocorrelation of the output signal is (Alencar, 2001)

$$Y(t) = \alpha X(t) - \beta X^2(t). \qquad (2.128)$$

$$
\begin{aligned}
R_Y(\tau) &= \mathrm{E}\left[Y(t) Y(t+\tau)\right] && (2.129) \\
&= \mathrm{E}\left[(\alpha X(t) - \beta X^2(t))(\alpha X(t+\tau) - \beta X^2(t+\tau))\right] \\
&= \alpha^2 R_X(\tau) - \alpha\beta \mathrm{E}[X(t) X^2(t+\tau) + X^2(t) X(t+\tau)] \\
&\quad + \beta^2 \mathrm{E}[X^2(t) X^2(t+\tau)].
\end{aligned}
$$

As previously stated, for a Gaussian zero-mean input signal, it is possible to use Price's theorem (Price, 1958), to compute the autocorrelation

$$
\begin{aligned}
\mathrm{E}[X(t)^2 X(t+\tau)^2] &= 4\int_0^{C_X} \mathrm{E}[X(t)X(t+\tau)] dC(X(t), X(t+\tau)) \\
&\quad + \mathrm{E}[X(t)] \cdot \mathrm{E}[X(t+\tau)], && (2.130)
\end{aligned}
$$

in which $C_X = C[X(t), X(t+\tau)]$ is the covariance between the variables $X(t)$ and $X(t+\tau)$.

The integral can be computed, giving as a result

$$R_Y(\tau) = \alpha^2 R_X(\tau) + 2\beta^2 R_X^2(\tau) + \beta^2 P_X^2. \tag{2.131}$$

The autocorrelation for $\tau = 0$ is the total power at the output of the non-linear amplifier,

$$
\begin{aligned}
P_Y &= R_Y(0) = \alpha^2 R_X(0) + 2\beta^2 R_X^2(0) + \beta^2 P_X^2 \\
&= \alpha^2 P_X + 2\beta^2 P_X^2 + \beta^2 P_X^2 \\
&= \alpha^2 P_X + 3\beta^2 P_X^2.
\end{aligned}
$$

The autocorrelation for $\tau \to \infty$ is the DC power at the output of the non-linear amplifier. Recalling that the input autocorrelation vanishes for $\tau \to \infty$,

$$P_Y = \beta^2 P_X^2.$$

Finally, the Wiener–Khintchin theorem permits the computation of the PSD of the non-linear amplifier output signal

$$S_Y(\omega) = \alpha^2 S_X(\omega) + \frac{2\beta^2}{2\pi} S_X(\omega) * S_X(\omega) + \beta^2 P_X^2 \delta(\omega). \tag{2.132}$$

The spectrum shows an amplified version of the input spectrum, an impulse at the origin, that corresponds to a DC level, in addition to a convolution of the input signal spectra. The convolution, induced by the polynomial in the second term, spreads the spectrum at the output and doubles the signal bandwidth.

2.7.5 Response of an Ideal Diode

The Shockley ideal diode equation relates the diode current $I(t)$ of a PN junction diode to the random voltage applied across the diode terminals, $(V(t))$. This relationship gives the diode characteristic,

$$I(t) = I_0 \left(e^{\frac{V(t)}{V_T}} - 1 \right), \tag{2.133}$$

in which I_0 is the reverse bias saturation current, the diode leakage current density in the absence of light, typically 10^{-12} A, and V_T is the thermal voltage, that is approximately 25.85 mV at ambient temperature (300 K).

In general, the thermal voltage V_T is a parameter defined as

$$V_T = \frac{kT}{q} \tag{2.134}$$

in which $k = 1.38064852 \times 10^23$ J/K is the Boltzmann constant, T is the absolute temperature of the PN junction, in kelvin (K), and the absolute value of the electron charge is $q = 1.6021766208 \times 10^{19}$ C.

The autocorrelation of the diode current is

$$R_I(\tau) = \mathrm{E}[I(t)I(t+\tau)] = I_0^2 \cdot \mathrm{E}\left[\left(e^{\frac{V(t)}{V_T}} - 1\right)\left(e^{\frac{V(t+\tau)}{V_T}} - 1\right)\right]. \tag{2.135}$$

This equation can be written as

$$R_I(\tau) = I_0^2 \cdot \mathrm{E}\left[e^{\frac{V(t)+V(t+\tau)}{V_T}} - e^{\frac{V(t)}{V_T}} - e^{\frac{V(t+\tau)}{V_T}} + 1\right],$$

and applying the linearity property of the expected value operator, one obtains

$$R_I(\tau) = I_0^2 \cdot \mathrm{E}\left[e^{\frac{V(t)+V(t+\tau)}{V_T}}\right] - I_0^2 \cdot \mathrm{E}\left[e^{\frac{V(t)}{V_T}}\right] - I_0^2 \cdot \mathrm{E}\left[e^{\frac{V(t+\tau)}{V_T}}\right] + I_0^2. \tag{2.136}$$

To solve the set of equations, first recall that it is possible to compute the the expected value of a stochastic function $f(X(t))$ of a stationary random process $X(t)$, when the probability distribution of $X(t)$ is known, but the distribution of $g(X(t))$ is not explicitly known.

$$\mathrm{E}\left[g(X)\right] = \int_{-\infty}^{\infty} g(x)p_X(x)dx \tag{2.137}$$

Assume that the voltage across the diode is a Gaussian process, with mean $\mathrm{E}\left[V(t)\right] = \mu$ and variance σ^2. Let $V(t) = \mu + \sigma Z(t)$, in which $Z(t)$ is a standard normal, with zero mean and unit variance, and drop the time independent variable for a while, considering that the input signal is stationary.

Then it is possible to write (Lévine, 1973),

$$\mathrm{E}\left[e^V\right] = \mathrm{E}\left[e^{\mu+\sigma Z}\right] = \int_{-\infty}^{\infty} e^{\mu+\sigma z}p_Z(x)dz. \tag{2.138}$$

Substituting the pdf for the stationary random process $Z(t)$, one obtains

$$\mathrm{E}\left[e^V\right] = \frac{1}{\sqrt{2\pi}} \int_{-\infty}^{\infty} e^{\mu+\sigma z}e^{-z^2/2}dz = \frac{e^\mu}{\sqrt{2\pi}} \int_{-\infty}^{\infty} e^{\sigma z}e^{-z^2/2}dz.$$

Completing the square of the expression $\sigma z - \frac{z^2}{2}$,

$$\frac{1}{2}\left(z^2 - 2\sigma z\right) = \frac{1}{2}(z - \sigma)^2 - \frac{1}{2}\sigma^2.$$

Therefore, the integral is

$$\mathrm{E}\left[e^V\right] = \frac{e^{\mu + \sigma^2/2}}{\sqrt{2\pi}} \int_{-\infty}^{\infty} e^{(z-\sigma)^2}\,dz = e^{\mu + \sigma^2/2}. \qquad (2.139)$$

Considering that the signal is stationary, $\mathrm{E}[X(t+\tau)] = \mathrm{E}[X(t)]$, and that the input signal has zero mean, that is, $\mu = 0$, to simplify the expression, and using Formula (2.139), then Equation (2.136) can be written as

$$R_I(\tau) = I_0^2 \cdot \exp\left(\frac{\mathrm{E}[(V(t) + V(t+\tau))^2]}{2V_T^2}\right) - 2I_0^2 \cdot \mathrm{E}\left[e^{\frac{\sigma^2}{2V_T^2}}\right] + I_0^2. \quad (2.140)$$

Expanding the square inside the exponential function, one obtains

$$R_I(\tau) = I_0^2 \cdot \exp\left(\frac{\mathrm{E}[V^2(t) + V^2(t+\tau) + 2V(t)V(t+\tau)]}{2V_T^2}\right)$$
$$- 2I_0^2 \cdot e^{\frac{\sigma^2}{2V_T^2}} + I_0^2. \qquad (2.141)$$

Computing the expected values inside the exponential function leads to

$$R_I(\tau) = I_0^2 \cdot \left[\exp\left(\frac{2\sigma^2 + R_V(\tau)}{2V_T^2}\right) - e^{\frac{\sigma^2}{2V_T^2}} + 1\right], \qquad (2.142)$$

in which $R_V(\tau) = \mathrm{E}[V(t)V(t+\tau)]$ is the autocorrelation of the input voltage across the diode.

As can be verified, in the limit as $\tau \to \infty$, the autocorrelation vanishes and Formula (2.142) gives the diode saturation power that is obtained if the reverse bias saturation current I_0 passes through a resistance of 1 Ω.

$$R_I(\infty) = I_0^2 \cdot \left[e^{\frac{2\sigma^2}{2V_T^2}} - e^{\frac{\sigma^2}{2V_T^2}} + 1\right] = I_0^2.$$

The diode current spectrum can be computed from Equation (2.142), but only for special cases. For example, the exponential can be expanded in a

Taylor series, and the first terms can be kept, to use the Wiener–Khintchin theorem.

$$R_I(\tau) = I_0^2 \cdot \left[1 + \frac{2\sigma^2 + R_V(\tau)}{2V_T^2} + \left(\frac{2\sigma^2 + R_V(\tau)}{2V_T^2}\right)^2 / 2 \cdots\right]$$
$$- I_0^2 \cdot \left[1 - e^{\frac{\sigma^2}{2V_T^2}}\right].$$

Expanding the binomial term, one obtains

$$R_I(\tau) = I_0^2 \cdot \left[1 + \frac{2\sigma^2 + R_V(\tau)}{2V_T^2}\right.$$
$$+ \frac{4\sigma^4 + R_V^2(\tau) + 4\sigma^2 R_V(\tau)}{8V_T^4} \cdots \right]$$
$$+ I_0^2 \cdot \left[1 - e^{\frac{\sigma^2}{2V_T^2}}\right].$$

Considering only the three first terms of the series expansion, and collecting the constant terms, one obtains a useful approximation of the output autocorrelation function,

$$R_I(\tau) \approx \frac{I_0^2}{8V_T^4} \cdot \left[4(V_T^2 + \sigma^2)R_V(\tau) + R_V^2(\tau)\right]$$
$$+ I_0^2 \cdot \left[2 - e^{\frac{\sigma^2}{2V_T^2}} + \frac{2\sigma^2}{2V_T^2} + \frac{4\sigma^4}{8V_T^4}\right].$$

The Wiener–Khintchin theorem gives the PSD of the diode current,

$$S_I(\omega) \approx \frac{I_0^2}{8V_T^4} \cdot \left[4(V_T^2 + \sigma^2)S_V(\omega) + \frac{S_V(\omega) * S_V(\omega)}{2\pi}\right]$$
$$+ 2\pi I_0^2 \cdot \left[2 - e^{\frac{\sigma^2}{2V_T^2}} + \frac{2\sigma^2}{2V_T^2} + \frac{4\sigma^4}{8V_T^4}\right]\delta(\omega).$$

As a consequence of the autocorrelation products that appear in the series expansion, the signal bandwidth of the output current increases, because of the resulting convolution terms in frequency. The impulse at the origin is caused by the diode constant current.

3

Amplitude Modulation Theory

3.1 Introduction

The design of systems to transmit reliable information, over non-reliable noisy and fading channels, is the basic objective of communications engineering (Alencar, 2015). Most communication systems, including the mobile cellular network, the television broadcast network, the telephony network, the satellite communication system, the radio broadcast network, and fiber optic transmission network, to name a few, rely on a basic mathematical model, as illustrated in the block diagram in Figure 3.1.

Information is carried by means of a modulated signal, using a propagation medium. For example, a propagation media, such as the optical fiber, is used to confine and guide the signals, but that is not always the case, since in wireless mobile communication, the propagation medium is the atmosphere, and satellite transmission uses the outer space.

When using radio transmitters, it would be impractical to transmit signals in the human voice frequency band directly since, by the theory of electromagnetic waves, efficient signal radiation can be achieved only with a radiating antenna size of the same order as the wavelength of the signals to be radiated. The radiating antennas would require sizes of hundreds of kilometers in order to radiate efficiently (Schwartz, 1970).

Therefore, to allow efficient radiation of radio signals, the communication systems employ carrier waves, whose parameters are compatible with the dimensions of the radiating element and with the media characteristics. Such electromagnetic waves provide the support to carry information (Alencar and Queiroz, 2010).

Figure 3.1   Transmission system model.

The information produced by the source is represented by an electrical modulating signal, used to modify one or more parameters of the carrier, which is modeled as a sinusoidal function.

As mentioned, modulation consists of the variation of one or more characteristics of the carrier waveform as a function of the modulating signal. The modulation is performed in three general ways.

- Amplitude Modulation (AM), if amplitude is the carrier parameter that is varied as a function of the modulating signal.
- Angle Modulation, if the carrier is Phase (PM) or Frequency Modulated (FM).
- Quadrature Amplitude Modulation (QAM), if both the amplitude and the phase of the carrier are varied simultaneously.

3.2 Amplitude Modulation

Amplitude modulation is a scheme in which the carrier wave has its instantaneous amplitude varied in accordance with the modulating signal (Alencar, 1999). Amplitude modulation, also known as Double-Sideband Amplitude Modulation (AM-DSB), is the most widely used modulation system and was adopted for commercial radio and television analog broadcast (Alencar, 2009a).

This is a consequence of some practical and economical advantages of the scheme such as, for example, simplicity of the receiver design and easy maintenance. The AM carrier is a sinusoid represented as (Alencar and da Rocha Jr., 2005).

$$c(t) = A\cos(\omega_c t + \phi), \tag{3.1}$$

in which A denotes the carrier amplitude, ω_c denotes the angular frequency (measured in radians per second, that is, rad/s), and ϕ denotes the carrier phase. Figure 3.2 illustrates an AM modulator.

The message signal, or modulating signal, usually denoted as $m(t)$, must be such that its maximum frequency is still much less than the carrier frequency, or $\omega_M \ll \omega_c$. The available modulation bandwidth is restricted by the linear region of operation of the transmitter power amplifiers, and varies from 0.1% to 1% of the carrier frequency (Gagliardi, 1978). The modulation process is illustrated in Figure 3.3.

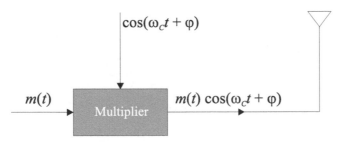

Figure 3.2 Illustration of an AM modulator.

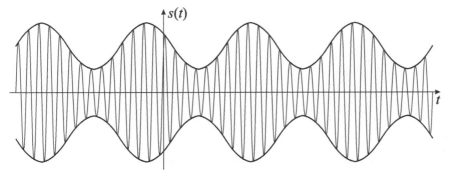

Figure 3.3 Example of amplitude modulation in the time domain.

In commercial broadcast, in order to have the carrier amplitude varying proportionally to $m(t)$, the instantaneous carrier amplitude has the form

$$a(t) = A + Bm(t) = A[1 + \Delta_{AM}m(t)], \tag{3.2}$$

in which $\Delta_{AM} = B/A$ is called the AM modulation index. The instantaneous amplitude is responsible for producing the modulated waveform as follows

$$s(t) = a(t)\cos(\omega_c t + \phi), \tag{3.3}$$

or

$$s(t) = A[1 + \Delta_{AM}m(t)]\cos(\omega_c t + \phi). \tag{3.4}$$

The effect of the modulating signal on the carrier is seen in Figure 3.4 that shows the modulating waveform, which alters the carrier amplitude to produce the modulated waveforms shown in Figures 3.5–3.7, for various values of the modulation index.

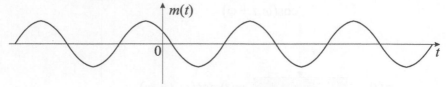

Figure 3.4   Modulating signal $m(t)$.

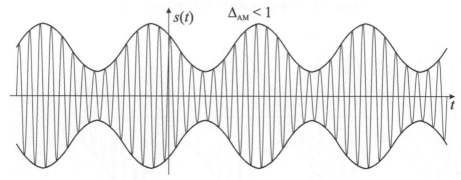

Figure 3.5   Effect of variation of the modulation index, for $\Delta_{AM} < 1$.

Figure 3.5 shows that the envelope of the modulated waveform has the same format as the modulating signal. This follows because $A + Bm(t) > 0$. In Figure 3.6 the modulation index is equal to 1, and the resulting signal is usually referred to as a 100% modulated carrier. In Figure 3.7, the modulation index is larger than 1, and causes phase inversion or phase rotation, in the modulated carrier, which is said to be over modulated.

The modulation index is an indication of how strong the modulating signal is with respect to the carrier. The modulation index should not exceed 100% in order to avoid distortion in a demodulated signal, if an envelope detector is employed. Depending on the manner by which the instantaneous amplitude varies as a function of Δ_{AM}, assuming $|m(t)| = 1$, the following terminology is used.

- $\Delta_{AM} = 1$ indicates 100% carrier modulation and allows envelope detection of the modulating signal.
- $\Delta_{AM} > 1$ indicates carrier overmodulation, causes phase rotation of the carrier, and requires synchronous demodulation to recover the modulating signal.

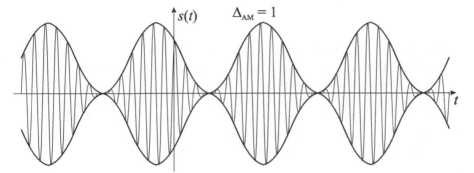

Figure 3.6 Effect of variation of the modulation index, for $\Delta_{AM} = 1$.

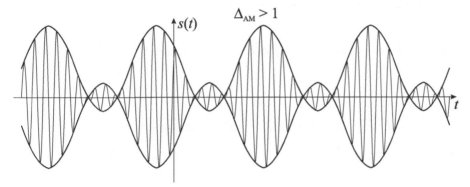

Figure 3.7 Effect of variation of the modulation index, for $\Delta_{AM} > 1$.

- $\Delta_{AM} < 1$ indicates carrier undermodulation and allows envelope detection of the modulating signal, but does not make efficient use of the carrier power.

3.3 Amplitude Modulation by Random Signals

The mathematical treatment used in this section involves concepts of random processes, or random signals. The derived results, which use an analysis based on random processes, compensate for the extra effort required to study the theory of stochastic processes. The theoretical development that follows is more elegant than the standard treatment, based on deterministic signals, and is also closer to real life signals.

Consider the modulated carrier given by

$$s(t) = a(t) \cos(\omega_c t + \phi), \tag{3.5}$$

in which

$$a(t) = A[1 + \Delta_{AM} m(t)].\tag{3.6}$$

The carrier phase ϕ is a random variable, with a uniform distribution in the interval $[0, 2\pi]$. The signal $m(t)$ is considered a stationary random process, with zero mean, and statistically independent of ϕ.

The carrier is a random process, because it has a random phase and because it was modulated by a stochastic process. The stationarity of the carrier is guaranteed, because the modulating signal is assumed to be stationary. The autocorrelation of a stationary random process is given by

$$R_S(\tau) = E[s(t)s(t+\tau)].\tag{3.7}$$

By substituting (3.3) into (3.7), it follows that

$$R_S(\tau) = E[(a(t)\cos(\omega_c t + \phi))(a(t+\tau)\cos(\omega_c(t+\tau) + \phi))],\tag{3.8}$$

or

$$R_S(\tau) = E[a(t)a(t+\tau)\cos(\omega_c t + \phi)\cos(\omega_c(t+\tau) + \phi)],\tag{3.9}$$

and by replacing the sum of cosines for a product of cosines, it follows that

$$R_S(\tau) = \frac{1}{2}E[a(t)a(t+\tau)(\cos(\omega_c\tau) + \cos(2\omega_c t + \omega_c\tau + 2\phi))].\tag{3.10}$$

Using the properties of the expectation operator, considering that $a(t)$ and ϕ are independent random variables, and that the mean value of the carrier is zero, one obtains,

$$R_S(\tau) = \frac{1}{2}R_A(\tau)\cos(\omega_c\tau),\tag{3.11}$$

in which the autocorrelation of the modulating signal $R_A(\tau)$ is defined as

$$R_A(\tau) = E[a(t)a(t+\tau)].\tag{3.12}$$

Replacing Expression (3.2) for $a(t)$ in (3.12), it follows that

$$
\begin{aligned}
R_A(\tau) &= E[A(1 + \Delta_{AM}m(t))A(1 + \Delta_{AM}m(t + \tau))] \qquad (3.13) \\
&= A^2 E[1 + \Delta_{AM}m(t) + \Delta_{AM}m(t + \tau) + \Delta_{AM}{}^2 m(t)m(t + \tau)].
\end{aligned}
$$

Again using properties of the expectancy, and recalling that $m(t)$ is a stationary process, with zero mean, that is, $E[m(t)] = E[m(t + \tau)] = 0$, it follows that

$$
R_A(\tau) = A^2[1 + \Delta_{AM}^2 R_M(\tau)], \qquad (3.14)
$$

in which $R_M = E[m(t)m(t + \tau)]$ represents the autocorrelation of the message signal. Finally, the autocorrelation of the amplitude-modulated carrier is given by

$$
R_S(\tau) = \frac{A^2}{2}[1 + \Delta_{AM}^2 R_M(\tau)] \cos \omega_c \tau. \qquad (3.15)
$$

3.3.1 Power of an AM Carrier

The power of an AM carrier is obtained as the value of its autocorrelation for $\tau = 0$, and therefore,

$$
P_S = R_S(0) = \frac{A^2}{2}(1 + \Delta_{AM}^2 P_M), \qquad (3.16)
$$

in which $P_M = R_M(0)$ represents the power in the message signal $m(t)$. The power in the non-modulated carrier is given by $\frac{A^2}{2}$, as can be verified, and represents a significant portion of the total transmitted power.

3.3.2 Power Spectral Density

As previously discussed, the PSD of the AM modulated carrier is obtained as the Fourier transform of the autocorrelation function $R_S(\tau)$. This result is known as the Wiener–Khintchin theorem,

$$
S_S(\omega) = \mathcal{F}[R_S(\tau)],
$$

in which $R_S(\tau) = \frac{1}{2} R_A(\tau) \cos(\omega_c \tau)$.

The autocorrelation function $R_S(\tau)$ can be seen as the product of two functions: $\frac{1}{2} R_A(\tau)$ and $\cos(\omega_c \tau)$. Using this line of reasoning, the Fourier transform of $R_s(\tau)$ is calculated with the application of the convolution theorem, that is,

$$S_S(\omega) = \mathcal{F}[R_S(\tau)] = \frac{1}{2\pi}\left(\mathcal{F}[\frac{1}{2}R_A(\tau)] * \mathcal{F}[\cos(\omega_c\tau)]\right)$$

$$= \frac{1}{2\pi}\left[\frac{1}{2}S_A(\omega) * (\pi\delta(\omega+\omega_c) + \pi\delta(\omega-\omega_c))\right],$$

and applying the impulse filtering property, it follows that

$$S_S(\omega) = \frac{1}{4}\left[S_A(\omega+\omega_c) + S_A(\omega-\omega_c)\right], \tag{3.17}$$

in which $S_A(\omega) = \mathcal{F}[R_A(\tau)]$.

The PSD of the modulating signal can be derived by writing the expression for $R_A(\tau)$, and then calculating its Fourier transform.

$$S_A(\omega) = \mathcal{F}[A^2(1 + \Delta_{AM}{}^2 R_M(\tau)],$$
$$= A^2[2\pi\delta(\omega) + \Delta_{AM}{}^2 S_M(\omega)].$$

in which $S_M(\omega)$ is the PSD of the message signal.

The PSD of the modulated AM carrier is finally given by

$$S_S(\omega) = \frac{A^2}{4}\left[2\pi(\delta(\omega+\omega_c) + \delta(\omega-\omega_c))\right.$$
$$\left. + \Delta_{AM}{}^2(S_M(\omega+\omega_c) + S_M(\omega-\omega_c))\right]. \tag{3.18}$$

The PSD of the message signal is sketched in Figure 3.8. Figure 3.9 illustrates the addition of a DC level to introduce the carrier. The modulated carrier is finally illustrated in Figure 3.10.

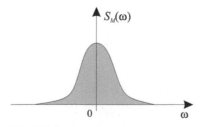

Figure 3.8   Original message signal to be transmitted.

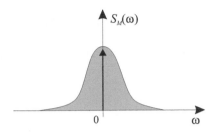

Figure 3.9 Power spectral density of signal plus DC level.

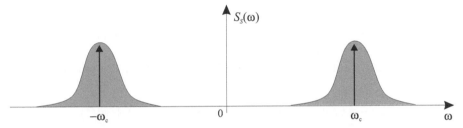

Figure 3.10 Power spectral density the modulated carrier.

The bandwidth required for the transmission of an AM signal is twice the bandwidth of the message signal. In commercial AM radio broadcast, the maximum frequency of the message signal is limited to 5 kHz, and consequently the AM bandwidth for commercial radio broadcast is 10 kHz.

3.4 Amplitude Modulators

Modulators are devices that allow, from a modulating signal and the carrier, to produce the modulated carrier waveform. In general, the AM modulators can be classified into two basic categories as follows.

1. Quadratic modulators.
2. Synchronous modulators.

3.4.1 Quadratic Modulator

Every nonlinear device having in its characteristic curve a term of degree two can be used as a modulator. In case the characteristic curve also contains terms of order higher than two, the undesirable effect of such terms may be filtered out at a subsequent stage. The quadratic modulator must possess an element of nonlinear characteristic containing predominantly a quadratic term. The Field Effect Transistor (FET) is a device that has this characteristic. The modulating circuit is shown in Figure 3.11.

Let $x(t)$ denote the FET input signal and let $y(t)$ denote the corresponding output signal associated with the FET. The FET operation curve can be described by the following equation

$$y(t) = \alpha x(t) - \beta x^2(t) \tag{3.19}$$

The input signal is the sum of carrier and the modulating signal,

$$x(t) = c(t) + m(t), \tag{3.20}$$

in which the function $c(t) = \cos(\omega_c t + \phi)$ represents the unmodulated carrier, and $m(t)$ represents the modulating signal.

Substituting Equation (3.20) into (3.19), the modulator has an output signal given by

$$y(t) = \alpha[A\cos(w_c t + \phi) + m(t)] - \beta[A\cos(w_c t + \phi) + m(t)]^2,$$

that is,

$$\begin{aligned} y(t) = {}& \alpha A \cos(\omega_c t + \phi) + am(t) - \beta A^2 \cos^2(\omega_c t + \phi) \\ & - 2\beta A m(t)\cos(\omega_c t + \phi) - \beta m^2(t). \end{aligned}$$

Filtering the signal $y(t)$ with a bandpass filter centered in the frequency ω_c, it follows that

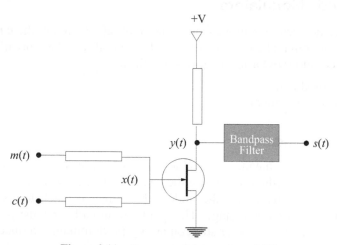

Figure 3.11   Square modulator using a FET.

$$s(t) = \alpha A \cos(\omega_c t + \phi) - 2\beta A m(t) \cos(\omega_c t + \phi)$$
$$= A[\alpha - 2\beta m(t)] \cos(\omega_c t + \phi), \tag{3.21}$$

which is the expression for the amplitude-modulated signal, with modulation index $\Delta_{AM} = 2\beta/\alpha$.

3.4.2 Synchronous Modulator

The synchronous modulator is based on the principle of sampling the signal resulting from the sum of the modulator signal and a DC level, and uses the properties of Fourier series.

The sampling operation is performed by means of a switching device that is synchronous with the carrier, as illustrated in Figure 3.12, in which PBF stands for Pass-Band Filter.

The sampled signal is $y(t) = x(t) \cdot f(t)$, in which $f(t)$ is a function of the switching operation and can be written as

$$f(t) = f_0 + \sum_{n=1}^{\infty} f_n \cos(n\omega_c t) \tag{3.22}$$

and $x(t) = B + m(t)$.

Therefore,

$$y(t) = [B + m(t)] \left[f_0 + f_1 \cos(\omega_c t) + \sum_{n=2}^{\infty} f_n \cos(n\omega_c t) \right]. \tag{3.23}$$

Filtering $y(t)$ with a bandpass filter centered in ω_c produces

$$s(t) = [B + m(t)] [f_1 \cos(\omega_c t)], \tag{3.24}$$

which is the expression for the Amplitude-Modulated Signal.

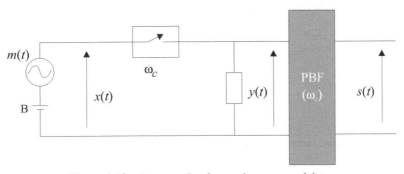

Figure 3.12   An example of a synchronous modulator.

3.4.3 Digital AM Signal

The digital amplitude-modulated signal, also called Amplitude Shift Keying (ASK), can be generated by the multiplication of the digital modulating signal by the carrier. The ASK signal is shown in Figure 3.13.

The ASK signal can be represented using a phase diagram, which consists in mapping the modulated signal in axes which are in phase (the I axis) and in quadrature (with a phase lag of $\pi/2$, the Q axis) with respect to the carrier phase.

This diagram is known as a constellation because it represents signal points as stars on a plane. The digital signal which amplitude modulates the carrier can be written as

$$m(t) = \sum_{k=-\infty}^{k=\infty} m_k p(t - kT_b), \qquad (3.25)$$

in which m_k represents the k-th randomly generated symbol, from a discrete alphabet, $p(t)$ is the pulse shape of the transmitted digital signal, and T_b is the bit interval.

The modulated signal without carrier is then given by

$$s(t) = \sum_{k=-\infty}^{\infty} m_k p(t - kT_b) \cos(\omega_c t + \phi). \qquad (3.26)$$

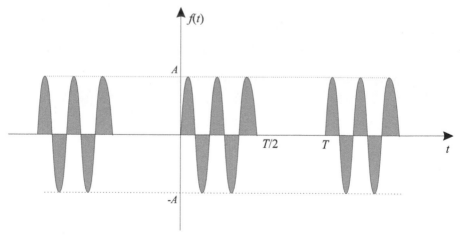

Figure 3.13   An example of a binary ASK signal.

As an example, Figure 3.14 shows the constellation diagram of a 4ASK signal, the symbols of which are $m_k \in \{-3A, -A, A, 3A\}$. All the signal points are on the cosine axis (in phase with the carrier), since there is no quadrature component in this case.

When the modulating signal is digital, the transmitted power is calculated considering the average power per symbol. For the ASK case, considering equiprobable symbols, it follows that the transmitted power is given by

$$P_S = \frac{1}{2} \sum_{k=1}^{4} m_k^2 p(m_k) = \frac{1}{2} \left[\frac{(-3A)^2 + (-A)^2 + (A)^2 + (3A)^2}{4} \right] = \frac{5}{2} A^2.$$

(3.27)

The probability of error for the coherent binary 2ASK is (Haykin, 1988)

$$P_e = \frac{1}{2} \text{erfc} \left(\sqrt{\frac{E_b}{N_0}} \right).$$

(3.28)

in which E_b is the binary pulse energy, N_0 represents the noise PSD and erfc$(\cdot)$ is the complementary error function,

$$\text{erfc}(x) = \frac{2}{\sqrt{\pi}} \int_x^\infty e^{-t^2} dt.$$

(3.29)

For a rectangular pulse, $E_b = A^2 T_b$, in which A is the pulse amplitude and T_b is the pulse duration.

3.4.4 AM Transmitter

Figure 3.15 shows a block diagram of an AM transmitter, normally used for radio broadcast. The transmitter consists of the following stages.

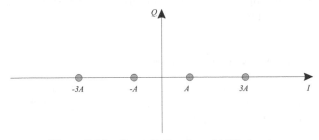

Figure 3.14   Constellation for a 4ASK signal.

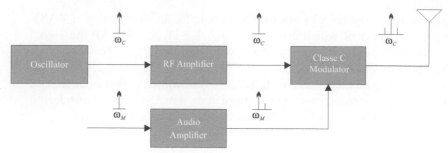

Figure 3.15 AM transmitter.

1. Oscillator, to generate the carrier.
2. Radiofrequency (RF) amplifier, to amplify the carrier power to a level adequate to feed the modulator stage.
3. Audio amplifier, to amplify the modulating signal.
4. Power (Class C) modulator, produces the AM signal when excited by the audio signal and by the carrier.

3.5 Suppressed Carrier Amplitude Modulation

In standard AM systems, most of the power is spent to transmit the carrier, that is a waste of power, because the carrier conveys no information.

In a Suppressed Carrier AM (AM-SC) system the carrier is not sent as part of the AM signal and all of the transmitter power is available for transmitting information over the two sidebands. The AM-SC system is more efficient than the AM system, regarding the use of transmitted power. However, a receiver for an AM-SC signal is more complex than an AM receiver.

The AM-SC technique translates the frequency spectrum of the modulating signal by multiplying it by a sinusoid, the frequency of which has a value equal to the desired frequency translation. In other words, the original message, or modulating, signal $m(t)$ becomes $m(t)\cos(\omega_c t + \phi)$, after multiplication by the carrier.

3.6 Spectrum of the AM-SC Signal

As before, let $s(t) = m(t)\cos(\omega_c t + \phi)$, in which $m(t)$ is a stationary random process with zero mean, ϕ is a random variable uniformly distributed in the interval $[0, 2\pi]$ and statistically independent of $m(t)$.

$$R_S(\tau) = \mathrm{E}[s(t)s(t+\tau)] \tag{3.30}$$

$$R_S(\tau) = \mathrm{E}\left[m(t)m(t+\tau)\cos(\omega_c t + \phi)\cos(\omega_c(t+\tau) + \phi)\right] \quad (3.31)$$

$$R_S(\tau) = \frac{1}{2}\mathrm{E}\left[m(t)m(t+\tau)\right]\cdot\mathrm{E}[\cos\omega_c\tau + \cos(\omega_c\tau + 2\omega_c t + 2\phi)] \quad (3.32)$$

$$R_S(\tau) = \frac{1}{2}\mathrm{E}\left[m(t)m(t+\tau)\right](\cos\omega_c\tau) \quad (3.33)$$

$$R_S(\tau) = \frac{R_M(\tau)}{2}\cos\omega_c(\tau) \quad (3.34)$$

The power of the AM-SC signal can be derived from the autocorrelation function computed for $\tau = 0$.

$$P_S = R_S(0) = \frac{P_M}{2}. \quad (3.35)$$

3.6.1 Power Spectral Density

The PSD of the modulated carrier is obtained by means of the Fourier transform of the autocorrelation function, as follows

$$S_S(\omega) = \int_{-\infty}^{\infty} R_S(\tau)e^{-j\omega\tau}\,d\tau, \quad (3.36)$$

$$S_S(\omega) = \frac{1}{4}S_M(\omega) * \delta[(\omega + \omega_c) + \delta(\omega - \omega_c)], \quad (3.37)$$

$$S_S(\omega) = \frac{1}{4}[S_M(\omega + \omega_c) + S_M(\omega - \omega_c)]. \quad (3.38)$$

3.6.2 The AM-SC Modulator

A common type of AM-SC scheme is the balanced modulator that uses a integrated circuit from the series MC1496, which includes a monolithic balanced modulator circuit, illustrated in Figure 3.16.

The device circuit consists of an upper quad differential amplifier driven by a standard differential amplifier with dual current sources. The output collectors are cross-coupled and, therefore, a balanced multiplication of the two input signals occurs (ON Semiconductor, 2018).

Because the output spectrum consists of the sum and difference of the two input frequencies, the device may be used, for example, as a balanced modulator, as a doubly balanced mixer, as a product detector, or as a frequency doubler. An application of the AM-SC modulator that uses the MC1496 is shown in Figure 3.17.

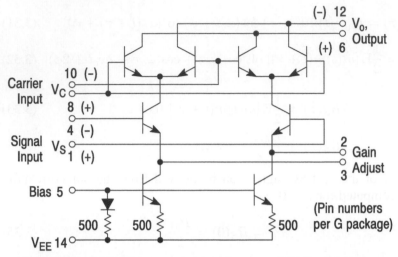

Figure 3.16   MC1496 integrated circuit. Courtesy of ON Semiconductor.

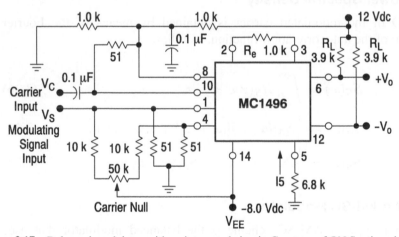

Figure 3.17   Balanced modulator with an integrated circuit. Courtesy of ON Semiconductor.

The circuit shown in Figure 3.17 may be used as an amplitude modulator; instead of the suppressed carrier operation, it is necessary to adjust only the carrier null potentiometer for the adequate amount of carrier insertion in the output signal.

3.7 AM-VSB Modulation

Amplitude Modulation Vestigial Sideband (AM-VSB) was used in the National Television Systems Committee (NTSC) and in the Phase Alternating Line (PAL) analog television standards. The Advanced Television Systems Committee (ATSC) also adopted AM-VSB for the American digital television standard.

The process of generating a Single Sideband (SSB) signal, explained in the next chapter, that makes a more efficient use of the available bandwidth, has some practical problems when performed using a sideband filter because in some applications a very sharp filter cutoff characteristic is required.

In order to overcome this design difficulty, a compromise solution was established between AM-DSB and SSB, called AM-VSB (Taub and Schilling, 1971). Figures 3.18 and 3.19 show the two types of modulations.

Figure 3.18 represents the spectrum of the modulating signal as well as the spectrum of the modulated AM-SC signal. Figure 3.19 presents the spectrum of a modulated SSB signal and the spectrum of a modulated AM-VSB signal.

As shown in Figure 3.19, the AM-VSB signal can be obtained by the partial suppression of the upper or lower sideband of an AM-SC. The AM-VSB scheme is used to transmit television video signals. Bandwidth reduction, practical implementation advantages, and lower cost are the main reasons for the choice of the AM-VSB system.

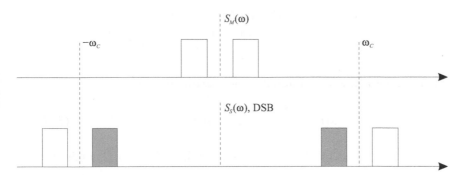

Figure 3.18 Spectra of the modulating signal $S_M(\omega)$ and of the AM-SC modulated carrier $S_S(\omega)$.

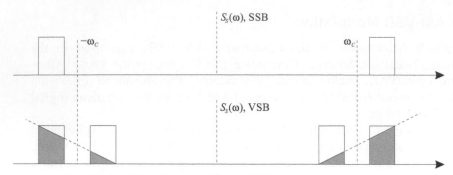

Figure 3.19 Spectra of SSB and VSB modulated signals.

3.8 Amplitude Demodulation

At the receiver, the modulated carrier can be synchronously demodulated to recover the message signal, as shown in Figure 3.20. The incoming signal

$$s(t) = A\left[1 + \Delta_{AM}m(t)\right]\cos(\omega_c t + \phi) \tag{3.39}$$

is multiplied (mixed) by a locally generated sinusoidal signal

$$l(t) = \cos(\omega_c t + \phi),$$

giving,

$$
\begin{aligned}
s(t) &= A\left[1 + \Delta_{AM}m(t)\right]\cos^2(\omega_c t + \phi) \\
&= A\left[1 + \Delta_{AM}m(t)\right]\left[\frac{1}{2} + \frac{1}{2}\cos(2\omega_c t + 2\phi)\right].
\end{aligned}
$$

Finally, the signal is low-pass filtered, which results in

$$s(t) = \frac{A}{2}\left[1 + \Delta_{AM}m(t)\right]. \tag{3.40}$$

The DC level then is blocked, to give the original signal $m(t)$ multiplied by a constant.

$$\tilde{m}(t) = \frac{m(t)}{2}.$$

If a problem occurs, and the phase of the local oscillator is not the same as the received signal phase,

$$l(t) = \cos(\omega_c t + \varphi), \tag{3.41}$$

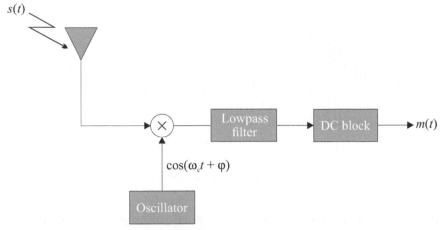

Figure 3.20 Block diagram of an AM demodulator.

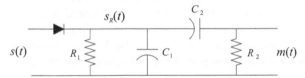

Figure 3.21 Example of a circuit for AM envelope demodulation.

the signal is attenuated, and fading occurs,

$$s(t) = A\left[1 + \Delta_{AM} m(t)\right]\left[\frac{1}{2}\cos(\phi - \varphi) + \frac{1}{2}\cos(2\omega_c t + \phi + \varphi)\right],$$

by a term proportional to the cosine of the phase difference, $\Delta\phi = \phi - \varphi$.

$$\tilde{m}(t) = \frac{m(t)}{2}\cos\Delta\phi. \tag{3.42}$$

If the frequency of the local oscillator is different from the frequency of the received carrier,

$$l(t) = \cos(\omega_l t + \phi),$$

the demodulated signal can remain frequency translated,

$$s(t) = A\left[1 + \Delta_{AM} m(t)\right]\left[\frac{1}{2}\cos(\omega_c - \omega_l)t + \frac{1}{2}\cos((\omega_c + \omega_l)t + \phi)\right],$$

to a residual frequency difference $\Delta\omega = \omega_c - \omega_l$, then,

$$\tilde{m}(t) = \frac{m(t)}{2}\cos\Delta\omega t. \tag{3.43}$$

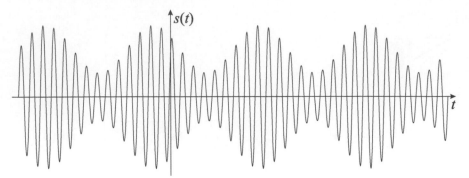

Figure 3.22   The modulated carrier at the input of the circuit.

The signal can also be recovered using an envelope detector, also called a non-coherent detector, that is simpler and inexpensive, as shown in Figure 3.21. The received carrier shown in Figure 3.22 is first rectified by the diode, producing the rectified carrier signal *v(t)* illustrated in Figure 3.23.

The rectified carrier is filtered by the low-pass filter formed by R_1 and C_1, to remove the remaining carrier. Finally, the DC component is blocked by capacitor C_2 to yield the message signal *m(t)*, sketched in Figure 3.24.

3.9 Noise Performance of Amplitude Modulation

When analyzing the noise performance of modulation systems, it is usual to employ the so-called quadrature representation for the noise. The quadrature representation for the noise $n(t)$ is presented in the following, as a function of its in phase $n_I(t)$ and quadrature $n_Q(t)$ components.

$$n(t) = n_I(t)\cos(\omega_c t + \phi) + n_Q(t)\sin(\omega_c t + \phi). \qquad (3.44)$$

The in phase and quadrature components have the same mean and variance as the noise $n(t)$, that is, $\sigma_I^2 = \sigma_Q^2 = \sigma^2$, in which σ^2 represents the noise

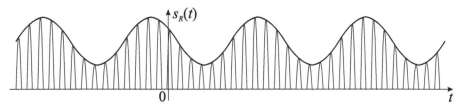

Figure 3.23   The modulated carrier after rectification.

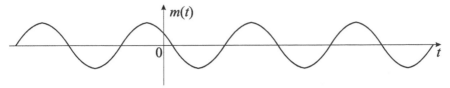

Figure 3.24 The demodulated signal.

variance $n(t)$. Assuming that an AM-SC was transmitted and was corrupted by additive noise, the received signal is given by $r(t)$, as follows.

$$r(t) = Bm(t)\cos(\omega_c t + \phi) + n(t). \tag{3.45}$$

The signal to noise ratio (SNR_I) at the demodulator input is given by

$$SNR_I = \frac{B^2/2}{\sigma^2}, \tag{3.46}$$

in which $\sigma^2 = \frac{2\omega_M N_0}{2\pi}$, for a modulating signal with highest frequency equal to ω_M. As a consequence of synchronous demodulation, the demodulated signal after filtering is given by

$$\hat{m}(t) = Bm(t) + n_I(t), \tag{3.47}$$

thus producing the following output signal to noise ratio

$$SNR_O = \frac{B^2}{\sigma^2}. \tag{3.48}$$

In this manner, the demodulation gain η for an AM-SC system is equal to

$$\eta = \frac{SNR_O}{SNR_I} = 2. \tag{3.49}$$

Therefore, AM-VSB modulation has a demodulating gain approximately equal to that of the AM-SC modulation. Following a similar procedure it can be shown that the demodulation gain of ordinary AM, employing coherent demodulation, is given by

$$\eta = \frac{2\Delta_{AM}^2 P_M}{1 + \Delta_{AM}^2 P_M} \tag{3.50}$$

It is noticed that the demodulation gain of ordinary AM is at least 3 dB inferior the corresponding AM-SC demodulation gain, eventually reaching up to 6 dB in practice (Lathi, 1989).

4

Quadrature Amplitude
Modulation Theory

The Quadrature Amplitude Modulation (QAM) is a versatile scheme that uses sine and cosine orthogonality properties to allow the transmission of two different signals in the same carrier, using a common bandwidth. The modulated signal occupies a bandwidth equivalent to the AM signal, which makes it attractive to design new modulation systems.

The QAM modulator, which also has the acronym QUAM, can be assembled using two DSB-SC modulators. In this type of modulation, the information is transmitted by both the carrier amplitude and phase.

4.1 Quadrature Modulation with Random Signals

Some recent digital modulation schemes are based on QAM and its variations. This is a direct result of the bandwidth economy obtained with the technique, but also due to the QAM two-dimensional constellation, that allows more freedom of choice to design the system.

The quadrature modulated signal $s(t)$ can be written as a composition of two modulating signals, in phase and quadrature,

$$s(t) = b(t)\cos(\omega_c t + \phi) + d(t)\sin(\omega_c t + \phi), \qquad (4.1)$$

in which the random modulating signals, $b(t)$ and $d(t)$, can be correlated or uncorrelated. The type of correlation function that relates them is responsible for the properties of the resulting carrier spectrum.

It is possible to write the transmitted signal in a manner that reveals the carrier amplitude and phase modulation,

$$s(t) = \sqrt{b^2(t) + d^2(t)}\cos\left[\omega_c t - \tan^{-1}\left(\frac{d(t)}{b(t)}\right) + \phi\right], \qquad (4.2)$$

in which the modulating signal, or amplitude resultant, can be expressed as

$$a(t) = \sqrt{b^2(t) + d^2(t)}, \tag{4.3}$$

and the phase resultant is

$$\theta(t) = -\tan^{-1}\left[\frac{d(t)}{b(t)}\right]. \tag{4.4}$$

The autocorrelation function for the quadrature modulated carrier can be computed, as usual, from the definition

$$R_S(\tau) = \mathrm{E}[s(t) \cdot s(t+\tau)]. \tag{4.5}$$

Substituting Equation (4.1) into (4.5), one obtains

$$\begin{aligned} R_S(\tau) = {} & \mathrm{E}\left[[b(t)\cos(\omega_c t + \phi) + d(t)\sin(\omega_c t + \phi)]\right. \\ & \cdot [b(t+\tau)\cos(\omega_c(t+\tau) + \phi) \\ & + \left. d(t+\tau)\sin(\omega_c(t+\tau) + \phi)]\right]. \end{aligned} \tag{4.6}$$

Expanding the product gives

$$\begin{aligned} R_S(\tau) = {} & \mathrm{E}[b(t)b(t+\tau)\cos(\omega_c t + \phi)\cos(\omega_c(t+\tau) + \phi) \\ & + d(t)d(t+\tau)\sin(\omega_c t + \phi)\sin mega_c(t+\tau) + \phi) \\ & + b(t)d(t+\tau)\cos(\omega_c t + \phi)\sin(\omega_c(t+\tau) + \phi) \\ & + b(t+\tau)d(t)\cos(\omega_c(t+\tau) + \phi)\sin(\omega_c t + \phi)]. \end{aligned} \tag{4.7}$$

Using trigonometric and expectancy properties, and collecting terms that represent known autocorrelation functions, it follows that

$$\begin{aligned} R_S(\tau) = {} & \frac{R_B(\tau)}{2}\cos\omega_c\tau + \frac{R_D(\tau)}{2}\cos\omega_c\tau \\ & + \frac{R_{DB}(\tau)}{2}\sin\omega_c\tau - \frac{R_{BD}(\tau)}{2}\sin\omega_c\tau. \end{aligned} \tag{4.8}$$

Which can be simplified to

$$\begin{aligned} R_S(\tau) = {} & \left[\frac{R_B(\tau) + R_D(\tau)}{2}\right]\cos\omega_c\tau \\ & + \left[\frac{R_{DB}(\tau) - R_{BD}(\tau)}{2}\right]\sin\omega_c\tau. \end{aligned} \tag{4.9}$$

The previous expression is useful in the design of QAM systems. The required physical property of the system can be inferred from the autocorrelation and correlation functions. This is demonstrated in the design of the Single Sideband (SSB) system.

The autocorrelation function can be written as

$$R_S(\tau) = \frac{R(\tau)}{2} \cos\left[\omega_c \tau + \theta(\tau)\right], \tag{4.10}$$

in which

$$R(\tau) = \sqrt{\left[R_B(\tau) + R_D(\tau)\right]^2 + \left[R_{DB}(\tau) - R_{BD}(\tau)\right]^2} \tag{4.11}$$

and

$$\theta(\tau) = -\tan^{-1}\left[\frac{R_{DB}(\tau) - R_{BD}(\tau)}{R_B(\tau) + R_D(\tau)}\right]. \tag{4.12}$$

But, according to the symmetry property of the autocorrelation function, the second term of Equation (4.9) must also be even, and therefore, the expression $R_{DB}(\tau) - R_{BD}(\tau)$ must be an odd function, because it is multiplied by a sine function.

Considering zero mean uncorrelated modulating signals,

$$R_{BD}(\tau) = \mathrm{E}[b(t)d(t + \tau)] = 0$$

and

$$R_{DB}(\tau) = \mathrm{E}[b(t + \tau)d(t)] = 0.$$

The resulting autocorrelation is then given by

$$R_S(\tau) = \frac{R_B(\tau)}{2} \cos\omega_c\tau + \frac{R_D(\tau)}{2} \cos\omega_c\tau. \tag{4.13}$$

The carrier power is given by the following formula

$$P_S = R_S(0) = \frac{P_B + P_D}{2}. \tag{4.14}$$

The power spectrum density is obtained by applying the Fourier transform to the autocorrelation function, the Wiener–Khintchin theorem, which gives

$$\begin{aligned}
S_S(\omega) &= \frac{1}{4}\left[S_B(\omega + \omega_c) + S_B(\omega - \omega_c) + S_D(\omega + \omega_c) + S_D(\omega - \omega_c)\right] \\
&= \frac{j}{4}\left[S_{BD}(\omega - \omega_c) + S_{BD}(\omega + \omega_c)\right. \\
&\quad \left. + S_{DB}(\omega + \omega_c) - S_{DB}(\omega - \omega_c)\right],
\end{aligned} \tag{4.15}$$

in which $S_B(\omega)$ and $S_D(\omega)$ represent the respective power spectrum densities for $b(t)$ and $d(t)$; $S_{BD}(\omega)$ is the cross-spectrum density between $b(t)$ and $d(t)$; $S_{DB}(\omega)$ is the cross-spectrum density between $d(t)$ and $b(t)$.

For uncorrelated signals, the previous formula can be simplified to

$$S_S(\omega) = \frac{1}{4}\left[S_B(\omega + \omega_c) + S_B(\omega - \omega_c)\right]$$
$$+ \frac{1}{4}\left[S_D(\omega + \omega_c) + S_D(\omega - \omega_c)\right]. \qquad (4.16)$$

The QAM scheme is shown in Figure 4.1. It is a very versatile device, and can be used to assemble several different modulators.

4.2 Single Sideband Modulation

Single Side-Band (SSB) is a type of quadrature amplitude modulation, that uses either the lower or upper AM side band for transmission. The SSB signal can be obtained by filtering out the undesired side band of an AM-SC signal, or by using properties of the Hilbert filter.

The main advantage of the SSB signal is an economy of bandwidth and power, compared to other modulation systems, but it needs frequency and phase synchronization to be recovered (Carlson, 1975). In order to understand the process of generating the SSB signal, the Hilbert transform is introduced in the following section.

4.2.1 Hilbert Transform

The Hilbert transform is named after the German mathematician David Hilbert (1862–1943), who made notable contributions to several areas of

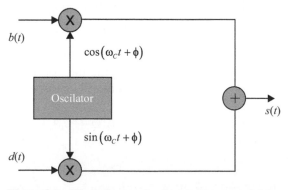

Figure 4.1   Block diagram for the quadrature modulator.

research, in Mathematics, formulated the theory of Hilbert spaces, gave the foundations of functional analysis, and also defended Georg Cantor's set theory.

The Hilbert transform maps a time function into another, in the same domain, what is called endomorfism. The Hilbert transform of a signal $f(t)$ is written $\hat{f}(t)$, and is defined as (Gagliardi, 1978)

$$\hat{f}(t) = \frac{1}{\pi} \int_{-\infty}^{\infty} \frac{f(\tau)}{t - \tau} d\tau. \tag{4.17}$$

The Hilbert transform is a linear operation and its inverse transform is given by

$$f(t) = -\frac{1}{\pi} \int_{-\infty}^{\infty} \frac{\hat{f}(t)}{t - \tau} d\tau. \tag{4.18}$$

Functions $f(t)$ and $\hat{f}(t)$ form a pair of Hilbert transforms. This transform shifts all frequency components of an input signal by $\pm 90^0$. The positive frequency components are shifted by -90^0, and the negative frequency components are shifted by $+90^0$. The spectral amplitudes are not affected.

From the definition, $\hat{f}(t)$ can be interpreted as the convolution of $f(t)$ and $\frac{1}{\pi t}$. The Fourier transform of the convolution of two signals is given by the product of the transforms of each signal. Therefore,

$$G(\omega) = \mathcal{F}[f(t) * \frac{1}{\pi t}] = \mathcal{F}[f(t)] \cdot \mathcal{F}[\frac{1}{\pi t}]. \tag{4.19}$$

4.2.2 Fourier Transform of $1/\pi t$

To obtain the Fourier transform of $\frac{1}{\pi t}$, one can resort to the following reasoning:

- Consider the signal function in Figure 4.2.
- The derivative of this function is an impulse with two units of area centered at the origin, whose Fourier transform is the constant function $\Delta(\omega) = 2$.
- Using the Fourier integral property,

$$\text{sgn}(t) = u(t) - u(-t) \leftrightarrow \frac{\Delta(\omega)}{j\omega} = \frac{2}{j\omega}.$$

- Finally, using the symmetry property, it follows that

$$\frac{1}{\pi t} \longleftrightarrow j \left[u(-\omega) - u(\omega) \right].$$

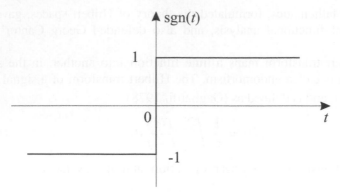

Figure 4.2   Signal function.

Therefore, one obtains

$$G(\omega) = j[u(-\omega) - u(\omega)] \cdot F(\omega). \tag{4.20}$$

4.2.3 Properties of the Hilbert Transform

The Hilbert transform puts the original function and its transform in quadrature. For example, the Hilbert transform of the cosine function is the sine. It has the following properties:

1. The signal $f(t)$ and its Hilbert transform $\hat{f}(t)$ have the same power spectrum density, that is, $S_{\hat{F}}(\omega) = S_F(\omega)$.
2. The signal $f(t)$ and its Hilbert transform $\hat{f}(t)$ have the same power autocorrelation, that is, $R_{\hat{F}}(\tau) = R_F(\tau)$.
3. The signal $f(t)$ and its Hilbert transform $\hat{f}(t)$ are orthogonal, that is, $R_{\hat{F}F}(\tau) = R_{F\hat{F}}(\tau) = 0$.
4. If $\hat{f}(t)$ is the Hilbert transform of $f(t)$, then the Hilbert transform of $\hat{f}(t)$ is $-f(t)$.

Figure 4.3 illustrates the Hilbert transform of a square wave.

4.2.4 Producing the SSB Signal

The SSB signal can be obtained by filtering one of the AM-SC sidebands, or by using the properties of the Hilbert transform. Consider a sinusoidal signal $m(t) = \cos(\omega_M)t$, which has a Fourier spectrum $S_M(\omega)$ represented by two impulses at $\pm\omega_M$.

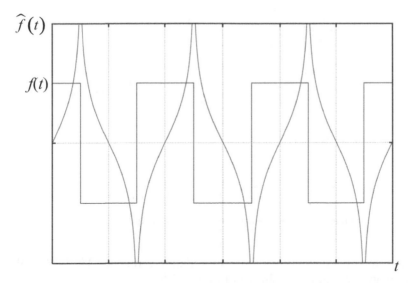

Figure 4.3 Hilbert transform of a square wave.

The modulated signal is given by $\cos(\omega_M t)\cos(\omega_c t)$, whose Fourier spectrum is that of $S_M(\omega)$ shifted to $\pm\omega_c$. The carrier spectrum is formed by two impulses at $\pm\omega_c$. Therefore, producing an SSB signal, for the special case, $m(t) = \cos(\omega_M t)$ is equivalent to generating the signal $\cos(\omega_c - \omega_M)t$, or the signal $\cos(\omega_c + \omega_M)t$.

By trigonometry, it follows that

$$\cos(\omega_c - \omega_M)t = \cos\omega_M t \cos\omega_c t + \sin\omega_M t \sin\omega_c t. \qquad (4.21)$$

Thus, the desired SSB signal can be produced adding $\cos\omega_M t \cos\omega_c t$ and $\sin\omega_M t \sin\omega_c t$. Signal $\cos\omega_M t \cos\omega_c t$ can be produced by a balanced modulator. Signal $\sin\omega_M t \sin\omega_c t$ can be written as $\cos(\omega_M t - \frac{\pi}{2})$ $\cos(\omega_c t - \frac{\pi}{2})$.

This signal can also be generated by a balanced modulator, as long as $\cos\omega_M t$ and carrier $\cos\omega_c t$ are phase shifted by $\frac{\pi}{2}$. Although this result has been derived for the special case $m(t) = \cos\omega_M t$, it is valid to any waveform, because of the Fourier series properties that allow the representation of any signal by a sum of sine and cosine functions.

The SSB associated with $m(t)$ is thus

$$s_{SSB}(t) = m(t)\cos(\omega_c t + \phi) + \hat{m}(t)\sin(\omega_c t + \phi), \qquad (4.22)$$

in which $\hat{m}(t)$ is obtained by phase shifting all frequency components of $m(t)$ by $\frac{\pi}{2}$. The block diagram of the modulator is shown in Figure 4.4.

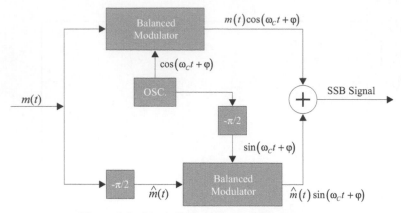

Figure 4.4   Block diagram for the SSB modulator.

A possible implementation of the modulator, using the MC1496/1596 series of integrated circuits, is shown in Figure 4.5.

4.2.5 Lower Sideband SSB with Random Signal

If the modulating signal $m(t)$ is a zero-mean, stationary, stochastic process, the usual procedure, to obtain the PSD, is to compute its autocorrelation function.

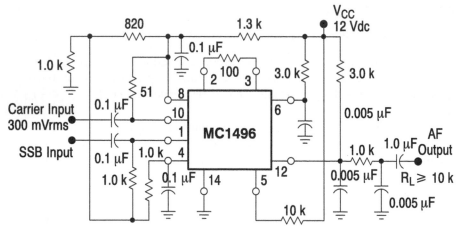

Figure 4.5   Implementation of the SSB modulator. Courtesy of ON Semiconductor.

The modulated SSB signal is given by

$$s_{SSB}(t) = m(t) \cos(\omega_c t + \phi) + \hat{m}(t) \sin(\omega_c t + \phi) \qquad (4.23)$$

The autocorrelation function is calculated, as usual, by the formula

$$R_S(\tau) = \mathrm{E}[s(t)s(t+\tau)] \qquad (4.24)$$

Substituting $s(t)$, given by Equation (4.23)

$$
\begin{aligned}
R_S(\tau) = {} & \mathrm{E}\left[(m(t) \cos(\omega_c t + \phi) + \hat{m}(t) \sin(\omega_c t + \phi))\right. \\
& \cdot (m(t+\tau) \cos(\omega_c(t+\tau) + \phi) \\
& \left. + \ \hat{m}(t+\tau) \sin(\omega_c(t+\tau) + \phi))\right].
\end{aligned}
\qquad (4.25)
$$

The previous equation can be split into

$$
\begin{aligned}
R_S(\tau) = {} & \mathrm{E}\left[m(t)m(t+\tau) \cos(\omega_c t + \phi) \cos(\omega_c(t+\tau) + \phi)\right] \\
& + \mathrm{E}\left[m(t)\hat{m}(t+\tau) \cos(\omega_c t + \phi) \sin(\omega_c(t+\tau) + \phi)\right] \\
& + \mathrm{E}\left[\hat{m}(t)m(t+\tau) \sin(\omega_c t + \phi) \cos(\omega_c(t+\tau) + \phi)\right] \\
& + \mathrm{E}\left[\hat{m}(t)\hat{m}(t+\tau) \sin(\omega_c t + \phi) \sin(\omega_c(t+\tau) + \phi)\right] \ (4.26)
\end{aligned}
$$

After the corresponding simplifications, one obtains

$$
\begin{aligned}
R_S(\tau) = {} & \frac{1}{2} R_{MM}(\tau) \cos \omega_c \tau + \frac{1}{2} R_{M\hat{M}}(\tau) \sin \omega_c \tau \\
& - \frac{1}{2} R_{\hat{M}M}(\tau) \sin \omega_c \tau + \frac{1}{2} R_{\hat{M}\hat{M}}(\tau) \cos \omega_c \tau.
\end{aligned}
\qquad (4.27)
$$

It is known that $R_{MM}(\tau) = R_{\hat{M}\hat{M}}(\tau)$ and $R_{M\hat{M}}(\tau) = -R_{\hat{M}M}(\tau)$. There-fore, the power spectrum density can be computed as $S_S(\omega) = \mathcal{F}[R_S(\tau)]$, by the use of previous relations and using the equation for the Hilbert filter

$$
H(\omega) = \begin{cases} -j & \text{se } \omega \geq 0 \\ +j & \text{se } \omega < 0 \end{cases}
\qquad (4.28)
$$

which leads to

$$
\begin{aligned}
S_{\hat{M}\hat{M}}(\omega) &= S_{MM}(\omega) & (4.29) \\
S_{M\hat{M}}(\omega) &= j[u(-\omega) - u(\omega)] \cdot S_{MM}(\omega) & (4.30) \\
S_{\hat{M}M}(\omega) &= j[u(\omega) - u(-\omega)] \cdot S_{MM}(\omega). & (4.31)
\end{aligned}
$$

Figure 4.6 illustrates the procedure.

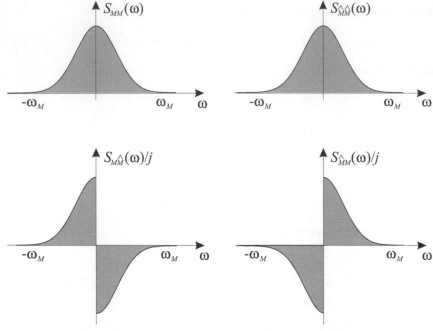

Figure 4.6   The initial part of single sideband modulation.

The following power spectral density results for the SSB signal

$$S(\omega)_{SSB} = S_M(\omega - \omega_c)u(-\omega + \omega_c) + S_M(\omega + \omega_c)u(\omega + \omega_c), \quad (4.32)$$

which represents the lower sideband SSB signal, obtained from the original spectrum $S_M(\omega)$.

Example: For the following SSB modulation scheme, determine the power of the modulated signal, considering that the modulating signal autocorrelation is given by $R_M(\tau) = \beta^2 e^{-\alpha|\tau|}$, and the modulated carrier is

$$s(t) = [1 + m(t)] \cos(\omega_c t + \phi) - \hat{m}(t) \sin(\omega_c t + \phi).$$

Solution: The power can be computed using the formula

$$P_S = \frac{P_A}{2},$$

in which $P_A = \mathrm{E}[a^2(t)]$ and

$$a(t) = \sqrt{(1 + m(t))^2 + \hat{m}^2(t)}$$

Considering that $m(t)$ has zero mean and that $P_M = P_{\hat{M}}$, it follows that

$$P_A = 1 + 2P_M.$$

Because $P_M = R_M(0) = \beta^2$, then

$$P_A = 1 + 2\beta^2 W$$

and

$$P_S = \frac{P_A}{2} = \frac{1 + \beta^2}{2} W. \tag{4.33}$$

4.3 ISB Modulation

Independent Side-Band (ISB) modulation permits sending two distinct signals in separate sidebands. The ISB signal can be obtained by the addition of two SSB signals, as shown in Figure 4.7.

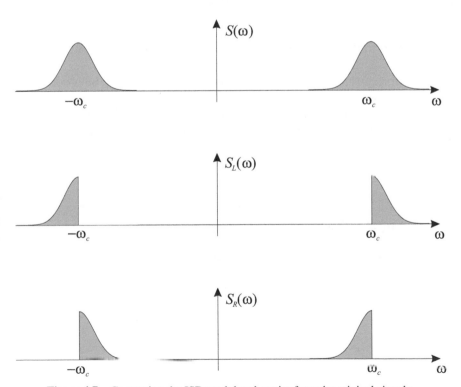

Figure 4.7 Generating the ISB modulated carrier from the original signals.

The expression for the upper sideband signal can be written as

$$s_L(t) = l(t)\cos(\omega_c t + \phi) - \hat{l}(t)\sin(\omega_c t + \phi) \qquad (4.34)$$

The expression for the lower sideband signal is given as

$$s_R(t) = r(t)\cos(\omega_c t + \phi) + \hat{r}(t)\sin(\omega_c t + \phi) \qquad (4.35)$$

The expression for the ISB signal is, then, the sum of the two signals

$$s(t) = s_L(t) + s_R(t) \qquad (4.36)$$

or

$$s(t) = [l(t) + r(t)]\cos(\omega_c t + \phi) + [\hat{r}(t) - \hat{l}(t)]\sin(\omega_c t + \phi). \qquad (4.37)$$

From the previous expression, it is noticed that the ISB signal can be produced using just one QAM modulator. To implement such a modulator, it is necessary to use a device to generate the sum and difference of the input signals.

The sum signal is applied to the in-phase modulator and the difference signal is applied to the quadrature modulator, after being phase shifted by a Hilbert filter.

The power spectrum density for the ISB modulated signal can be observed in Figure 4.7.

Example: Compute the power spectrum density and the ISB modulated signal average power, assuming the following power spectrum densities for the modulating signals

$$S_L(\omega) = S_0[u(\omega + \omega_M) - u(\omega - \omega_M)]$$

and

$$S_R(\omega) = S_0(1 - \frac{|\omega|}{\omega_M})[u(\omega + \omega_M) - u(\omega - \omega_M)].$$

Solution: The modulated signal power can be computed by

$$P_S = P_L + P_R$$

in which

$$P_L = \frac{1}{2\pi}\int_{-\infty}^{\infty} S_L(\omega)d\omega$$

and

$$P_R = \frac{1}{2\pi}\int_{-\infty}^{\infty} S_R(\omega)d\omega.$$

It is also possible to calculate the average power by integrating the power spectrum density function. Therefore, it follows that

$$P_L = \frac{\omega_M S_0}{\pi}$$

and

$$P_R = \frac{\omega_M S_0}{2\pi}.$$

Then,

$$P_S = \frac{3}{2}\frac{\omega_M S_0}{\pi} \text{ W}.$$

4.4 AM-Stereo

The industry promoted five main radio systems that use the AM-stereo technique. The common modulation technique is QUAM, but commercially it is called AM-stereo. The systems were introduced by the following companies: Belar, Magnavox, Kahn, Harris, and Motorola.

The Compatible QUAM (C-QUAM) was developed by Motorola as an AM-stereo system, with the objective of being compatible with the legacy AM mono receivers. The mono receivers should be able to detect and reproduce the sum of the stereo channels.

The stereo signal has a left channel l and a right channel r. The compatible AM-stereo envelope is proportional to the sum $l+r$ and its phase is a function of the difference signal $l - r$. Of course, the sum signal must be limited to 100% modulation index, at least for the negative peaks; otherwise, the phase modulation is lost.

Figure 4.8 shows the block diagram of the driver circuit to the system. The crystal oscillator, which produces the carrier, is usual in conventional AM transmitters. The driver circuit has a phase-shifter to generate a second carrier, which lags the first one by 90°.

The audio program channels are input to a matrix circuit, that is built with operational amplifiers, to produce the sum $l + r$ and difference signals $l - r$ which modulate the in-phase and quadrature carriers, respectively, using balanced modulators.

A low-level 25-Hz pilot signal that indicates the stereo operation is added to the $l - r$ signal, before modulation. The sidebands that come out of the modulators are added to the carrier, producing a QUAM signal, which is both amplitude and angle modulated.

A clipper circuit removes the amplitude modulation, leaving only the angle modulation, making the signal suitable to be further modulated. This

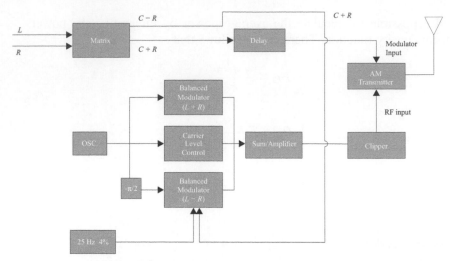

Figure 4.8   Block diagram of C-QUAM modulator.

signal is the fed, as a carrier, to a conventional amplitude modulator. The composite expression for the C-QUAM signal is

$$s(t)_{AM-EST} = A[1 + \Delta_B b(t)]\cos(\omega_c t + \theta(t)), \qquad (4.38)$$

in which

$$\theta(t) = \arctan\left[\frac{\Delta_D d(t) + \sin\omega_P t}{1 + \Delta_B b(t)}\right], \qquad (4.39)$$

in which $b(t) = l(t) + r(t)$ and $d(t) = l(t) - r(t)$.

4.5 Quadrature Amplitude Demodulation

The QUAM scheme requires synchronization, which can either be obtained with the transmission of a pilot signal, or by extracting the carrier from the modulated signal. At the receiver end, the modulated carrier is synchronously demodulated to recover the message signal, as shown in Figure 4.9. The incoming signal

$$s(t) = b(t)\cos(\omega_c t + \phi) + d(t)\sin(\omega_c t + \phi) \qquad (4.40)$$

is multiplied by two locally generated sinusoidal signals $\cos(\omega_c t + \phi)$ and $\sin(\omega_c t + \phi)$ and low-pass filtered. The operation recovers the original $b(t)$ and $d(t)$ signals, multiplied by a constant.

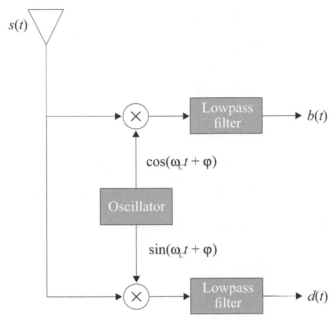

Figure 4.9 Block diagram of a QAM demodulator.

At the receiver, fading can occur if the local oscillator phase is different from the received carrier phase. The signal is attenuated by a term proportional to the cosine of the phase difference. As in the synchronous AM demodulation, a frequency drift of the local oscillator can shift the demodulated signal frequency and disturb the reception.

4.6 Performance Evaluation of SSB

The performance of the SSB modulator can be evaluated by computing the efficiency of the scheme at the receiving side. The SSB modulated signal, presented previously, can be written as

$$s(t) = m(t)\cos(\omega_c t + \phi) + \hat{m}(t)\sin(\omega_c t + \phi), \qquad (4.41)$$

in which $m(t)$ represents the message signal and $\hat{m}(t)$ its Hilbert transform.

The SSB signal spectral analysis shows that its bandwidth is the same as that of the baseband signal (ω_M). The received signal is expressed as

$$r(t) = [m(t) + n_I(t)]\cos(\omega_c t + \phi) + [\hat{m}(t) + n_Q(t)]\sin(\omega_c t + \phi). \qquad (4.42)$$

Because the noise occupies a bandwidth ω_M and demodulation is synchronous, the demodulation gain η for the SSB signal is the same as the one obtained for the AM-SC, that is,

$$\eta = \frac{SNR_O}{SNR_I} = 2. \tag{4.43}$$

4.7 Digital Quadrature Modulation

Most of the new communication systems are based on quadrature modulation. They use digital signals and are called QAM. The computation of the autocorrelation function and the power spectrum density follows the same rules previously established.

The modulating signals for the QAM scheme are

$$b(t) = \sum_{n=-\infty}^{\infty} b_n p(t - nT_b) \tag{4.44}$$

and

$$d(t) = \sum_{n=-\infty}^{\infty} d_n p(t - nT_b). \tag{4.45}$$

The modulated carrier is similar to the analog quadrature modulation,

$$s(t)_{QAM} = b(t)\cos(\omega_c t + \phi) + d(t)\sin(\omega_c t + \phi). \tag{4.46}$$

Figure 4.10 shows the constellation diagram for a 4-QAM signal, in which the symbols $b_n \in \{A, -A\}$ and $d_n \in \{A, -A\}$. It can be shown, using the same previous methods, that the modulated signal power for this special case is given by $P_S = A^2/2$.

Some authors call the signal Quadrature Phase Shift Keying, $\pi/4$-QPSK. It is used in schemes for mobile cellular communication systems. If the constellation is rotated by $\pi/4$, it produces another QAM modulation scheme, which presents more ripple than the previous one, because one of the carriers is shut off whenever the other carrier is transmitted.

Figure 4.11 shows the constellation diagram for a 16-QAM signal, whose points have coordinates $b_n = \{-3A, -A, A, 3A\}$, for the abscissa, and $d_n = \{-3A, -A, A, 3A\}$, for the ordinate.

The efficient occupation of the signal space renders the QAM technique more efficient than Amplitude Shift Keying (ASK) and Phase Shift Keying (PSK), in terms of bit error probability versus transmission rate. On the other hand, the scheme is vulnerable to non-linear distortion, which occurs

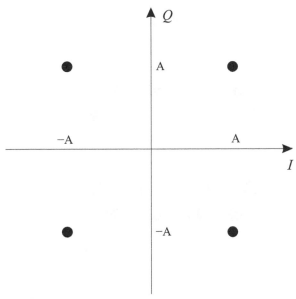

Figure 4.10 Constellation diagram for the 4-QAM signal.

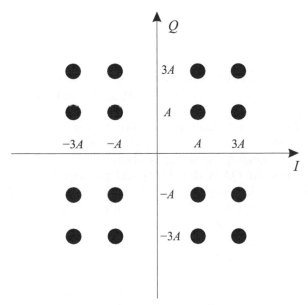

Figure 4.11 Constellation diagram for the 16-QAM signal.

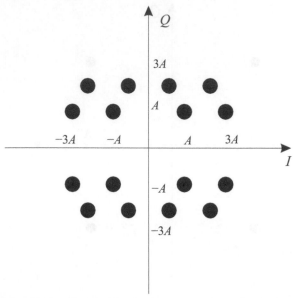

Figure 4.12   Constellation diagram for the 16-QAM signal, subject to distortion caused by a non-linear channel.

in power amplifiers on board satellite systems as shown, for example, in Figure 4.12.

Other problems can occur to degrade the transmitted signal. It is important that the communication engineer could identify them, to establish the best mitigation strategy.

Figure 4.13 shows a constellation diagram in which the symbols are contaminated by additive Gaussian noise. The circles illustrate, ideally, the uncertainty region surrounding the signal symbols.

Figure 4.14 shows the QAM signal affected by amplitude fading. The main fading effect is to decrease the distance between the symbols, which implies an increase in the Bit Error Rate (BER) of the system.

The symbol error probability is, usually, controlled by the smaller distance between the constellation symbols. The formulas for the error probability produce curves that decrease exponentially, or quasi-exponentially, with the signal to noise ratio.

The random phase variation effect, or jitter, is shown in Figure 4.15. This occurs when the local synchronization system is not able to perfectly follow the received signal phase variations. This introduces randomness in the detection process, which increases the bit symbol error probability.

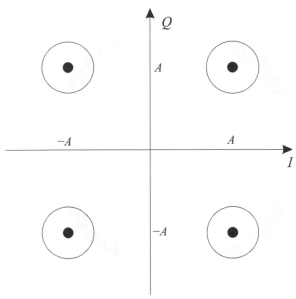

Figure 4.13 Constellation diagram for a 4-QAM signal with additive noise.

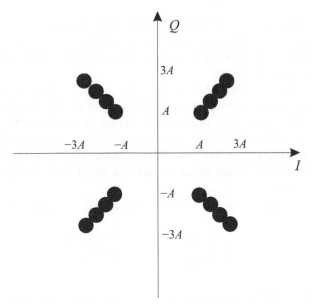

Figure 4.14 Constellation diagram for the 4-QAM signal, subject to fading.

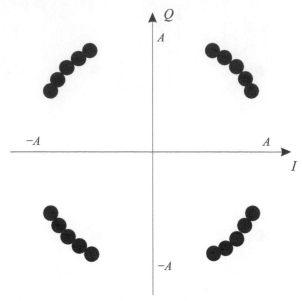

Figure 4.15   Constellation diagram for a 4-QAM signal, subject to jitter.

Figure 4.16 illustrates a 4-QAM signal affected by a total loss of synchronization. The constellation diagram seems to be a circle because the symbols rotate rapidly. In this case, detection is impossible, and the system, usually, has to be manually synchronized.

Figure 4.17 shows a constellation diagram in which the symbols are contaminated by additive Gaussian noise (El-Tanany et al., 2001). The circles illustrate, ideally, the uncertainty region surrounding the signal symbols.

The symbol error probability is, usually, controlled by the smaller distance between the constellation symbols. The formulas for the error probability produce curves that decrease exponentially, or quasi-exponentially, with the signal to noise ratio.

The probability of error for the 4-QAM signal is (Haykin, 1988)

$$P_e = \text{erfc}\left(\sqrt{\frac{E_b}{N_0}}\right) - \frac{1}{4}\text{erfc}^2\left(\sqrt{\frac{E_b}{N_0}}\right), \qquad (4.47)$$

which can be simplified to

$$P_e \approx \text{erfc}\left(\sqrt{\frac{E_b}{N_0}}\right), \qquad (4.48)$$

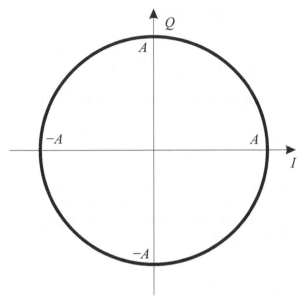

Figure 4.16 Constellation diagram for the 4-QAM, subject to synchronization loss.

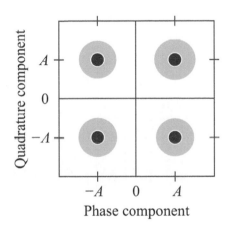

Figure 4.17 Constellation diagram for a 4-QAM signal with additive noise.

in which E_b is the pulse energy, N_0 represents the noise PSD, and erfc$(\cdot)$ is the complementary error function, defined as,

$$\text{erfc}(x) - \frac{2}{\sqrt{\pi}} \int_x^\infty e^{-t^2} dt. \tag{4.49}$$

For M symbols, the error probability is given by

$$P_e = 1 - \left[1 - \left(1 - \frac{1}{\sqrt{M}} \right) \text{erfc} \left(\sqrt{\frac{E_b}{N_0}} \right) \right]^2. \qquad (4.50)$$

For most purposes, the formula can be approximated by (Haykin, 1988).

$$P_e \approx 2 \left(1 - \frac{1}{\sqrt{M}} \right) \text{erfc} \left(\sqrt{\frac{E_b}{N_0}} \right). \qquad (4.51)$$

As discussed in this chapter, the stochastic analysis provides general expressions for the autocorrelation and PSD functions. The resulting PSD can be used to aid the design of modulation schemes, or to solve practical problems with the aid of spectrum analyzers.

5

Angle Modulation Theory

5.1 Introduction

Angle modulation is an expression that include both Frequency Modulation (FM) and Phase Modulation (PM), both analog and digital. Frequency modulation has an interesting history, because it was originally thought as a method to reduce the bandwidth required for the transmission of an audio signal. At the time, it was already known that amplitude modulation required twice the signal bandwidth for transmission.

John R. Carson was the first to recognize the problem with that argument. He explained the subject in a paper published in 1922 (Carson, 1922). But he failed, at that time, to appreciate the advantages of frequency over amplitude modulation, and the subject was forgotten for a while, until the engineer Edwin H. Armstrong invented the first equipment to frequency modulate a carrier, in 1931 (Armstrong, 1936).

Later on, Carson established, in an unpublished memorandum of August 28, 1939, that the bandwidth was equal to twice the sum of the peak frequency deviation, Δf, and the highest frequency of the modulating signal f_M (Carlson, 1975). This is the well-known Carson's rule,

$$B = 2(\Delta f + f_M). \tag{5.1}$$

Currently, several communication systems use either frequency or phase modulation, including analog and digital mobile cellular communication systems, satellite transmission systems, wireless telephones, and radio and television broadcast systems. The objective of this chapter is to present a general mathematical model to analyze stochastic angle-modulated signals.

5.2 Angle Modulation with Stochastic Signals

Most books present deterministic analyses of angle modulation, but it is well known that only the random signals matter being transmitted, because the

143

deterministic ones do not carry information. A deterministic waveform is only useful to test the transmission equipment.

This section presents a general and elegant method to compute the autocorrelation function and the power spectrum density of angle-modulated signals. Without loss of generality, the modulating signal is considered a zero-mean stationary stochastic process $m(t)$, with autocorrelation function $R_M(\tau)$. This is a common assumption, and a practical necessity.

5.2.1 Mathematical Model

Angle modulation is a non-linear operation, different from amplitude and quadrature modulations. Therefore, it requires a distinct line of analysis, which relies on approximating the equations using known formulas and adequate series. The following mathematical derivation provides an estimation of the spectrum of an angle-modulated signal, based on the first-order pdf of the stochastic modulating signal. The proof uses several properties of stochastic processes, and uses the linear mean square estimator (Papoulis, 1981).

In a general way, the angle-modulated carrier $s(t)$ is obtained from the following equations:

$$s(t) = A \cos\left(\omega_c t + \theta(t) + \phi\right), \tag{5.2}$$

in which

$$\theta(t) = \Delta_{FM} \cdot \int_{-\infty}^{t} m(t)dt, \tag{5.3}$$

for frequency modulation, and

$$\theta(t) = \Delta_{PM} \cdot m(t)dt, \tag{5.4}$$

for phase modulation, in which the constant parameters A, $\omega_c = 2\pi f_c$, Δ_{FM}, and Δ_{PM} represent, respectively, the carrier amplitude and angular frequency, and the frequency and phase deviation indices. The message signal is represented by $m(t)$, which is considered a zero-mean random stationary process.

The carrier phase ϕ is random, uniformly distributed in the range $(0, 2\pi]$, and statistically independent of $m(t)$. These are usual assumptions, considering that the phase cannot be determined *a priori*, and the uniform distribution maximizes the entropy, that is, the uncertainty regarding the phase information (Alencar, 2015).

The modulating signal, $\theta(t)$, represents the variation in the carrier phase, produced by the message. The frequency modulation index is defined as

$$\beta = \frac{\Delta_{FM}\sigma_M}{\omega_M}, \tag{5.5}$$

in which $\sigma_M = \sqrt{P_M}$, P_M represents the power of the message signal $m(t)$, and $\omega_M = 2\pi f_M$ is the maximum angular frequency of the signal.

The frequency deviation $\sigma_F = \Delta_{FM}\sigma_M$ represents the mean shift from the original, or spectral, carrier frequency ω_c. The modulation index gives and idea of how many times the modulating signal bandwidth fits into the frequency deviation.

The phase modulation index is defined as

$$\alpha = \Delta_{PM}\sigma_M. \tag{5.6}$$

The following steps are observed in the evaluation of the spectrum of the angle modulated signal:

1. Compute the autocorrelation function of the modulated carrier $s(t)$, from Equation (5.2).
2. Obtain an estimate of this autocorrelation, for the case of low, medium and high modulation indices, using the linear mean square estimator (Papoulis, 1981).
3. Compute the PSD of $s(t)$, using the Wiener–Khintchin theorem, as the Fourier transform of the signal autocorrelation.

The autocorrelation function of the modulated carrier, $s(t)$, defined by Equation (5.2), is expressed as

$$R_S(\tau) = \mathrm{E}[s(t)s(t+\tau)] \tag{5.7}$$

in which $\mathrm{E}[\cdot]$ represents the expected value operator.

Substituting the equation for the modulated carrier, and realizing that the expected value of a constant is the constant itself, one obtains

$$R_S(\tau) = A^2\mathrm{E}[\cos{(\omega_c t + \theta(t) + \phi)} \cdot \cos{(\omega_c(t+\tau) + \theta(t+\tau) + \phi)}].$$

Applying trigonometric properties, it follows that

$$R_S(\tau) = \frac{A^2}{2}\mathrm{E}[\cos{(\omega_c\tau - \theta(t) + \theta(t+\tau))} + \cos{(2\omega_c t + \omega_c\tau + \theta(t) + \theta(t+\tau) + 2\phi)}].$$

Because the expected value is a linear operator,

$$R_S(\tau) = \frac{A^2}{2}\mathrm{E}[\cos{(\omega_c\tau - \theta(t) + \theta(t+\tau))}]$$
$$+ \frac{A^2}{2}\mathrm{E}[\cos{(2\omega_c t + \omega_c\tau + \theta(t) + \theta(t+\tau) + 2\phi)}].$$

Again, using trigonometric properties to separate the random functions of time from the random variable, and considering that the phase is independent of the signal, one obtains

$$
\begin{aligned}
R_S(\tau) &= \frac{A^2}{2}\mathrm{E}\left[\cos\left(\omega_c\tau - \theta(t) + \theta(t+\tau)\right)\right] \\
&+ \frac{A^2}{2}\mathrm{E}\left[\cos\left(2\omega_c t + \omega_c\tau + \theta(t) + \theta(t+\tau)\right)\right] \cdot \mathrm{E}\left[\cos(2\phi)\right]. \\
&- \frac{A^2}{2}\mathrm{E}\left[\sin\left(2\omega_c t + \omega_c\tau + \theta(t) + \theta(t+\tau)\right)\right] \cdot \mathrm{E}\left[\sin(2\phi)\right].
\end{aligned}
$$

As can be verified, the expected value of the phase is zero, and therefore, the autocorrelation of the modulated carrier is

$$
R_S(\tau) = \frac{A^2}{2}\mathrm{E}[\cos(\omega_c\tau - \theta(t) + \theta(t+\tau))]. \tag{5.8}
$$

The computation of the modulated signal PSD is separated into three important cases, which are discussed in the following.

5.2.2 Low Modulation Index

This occurs when the modulation index is kept below a certain level, such as $\beta < 0.1$, which is usually obtained by controlling the modulating signal power at the input of the angle modulator, or by adjusting the frequency or phase deviations.

Recall that $\theta(t)$ is a function of $m(t)$, and that a small signal power P_M implies a small angle. The autocorrelation function of $s(t)$ can be obtained from Equation (5.8) by expanding the cosine function,

$$
\begin{aligned}
R_S(\tau) &= \frac{A^2}{2}\cos(\omega_c\tau)\mathrm{E}[\cos(-\theta(t) + \theta(t+\tau))] \\
&- \frac{A^2}{2}\sin(\omega_c\tau)\mathrm{E}[\sin(-\theta(t) + \theta(t+\tau))]. \tag{5.9}
\end{aligned}
$$

For a low modulation index, it is possible to expand the sine and cosine functions of Equation (5.9) in Taylor series, neglecting the high-order terms, which are only a small fraction of the first ones, to obtain

$$
\begin{aligned}
R_S(\tau) &= \frac{A^2}{2}\cos(\omega_c\tau)\mathrm{E}\left[1 - \frac{(-\theta(t) + \theta(t+\tau))^2}{2}\right] \\
&- \frac{A^2}{2}\sin(\omega_c\tau)\mathrm{E}\left[-\theta(t) + \theta(t+\tau)\right]. \tag{5.10}
\end{aligned}
$$

Considering that $m(t)$ is a zero-mean, stationary process, it follows that

$$R_S(\tau) = \frac{A^2}{2} \cos(\omega_c \tau)[1 - R_\Theta(0) + R_\Theta(\tau)], \qquad (5.11)$$

in which

$$R_\Theta(\tau) = \mathrm{E}[\theta(t)\theta(t+\tau)] \qquad (5.12)$$

and $R_\Theta(0) = P_\Theta$ is the power of the modulating signal $\theta(t)$.

The PSD of $s(t)$, for the FM case, is obtained by the use of the Wiener–Khintchin theorem, computing the Fourier transform of Equation (5.11) (Papoulis, 1983b), considering that for frequency modulation,

$$\theta(t) = \Delta_{FM} \cdot \int_{-\infty}^{t} m(t)dt.$$

Recalling that integration is a linear operation, and that

$$S_\Theta(\omega) = |H(\omega)|^2 S_M(\omega), \qquad (5.13)$$

therefore

$$S_\Theta(\omega) = \Delta_{FM}^2 \frac{S_M(\omega)}{\omega^2}, \qquad (5.14)$$

and the frequency-modulated carrier power spectrum density is given by

$$\begin{aligned} S_S(\omega) &= \frac{\pi A^2(1 - P_\Theta)}{2}[\delta(\omega + \omega_0) + \delta(\omega - \omega_0)] \\ &+ \frac{\Delta_{FM}^2 A^2}{4}\left[\frac{S_M(\omega + \omega_c)}{(\omega + \omega_c)^2} + \frac{S_M(\omega - \omega_c)}{(\omega - \omega_c)^2}\right], \qquad (5.15) \end{aligned}$$

in which $S_M(\omega)$ represents the PSD of the message signal $m(t)$, which has bandwidth ω_M. The modulated signal bandwidth is then double the message signal bandwidth BW $= 2\omega_M$.

From Equation (5.15), one can notice that the FM spectrum has the shape of the message signal spectrum multiplied by a squared hyperbolic function. The examples which are given in the following sections consider a uniform power spectrum density for the message signal.

The power of the angle excitation signal is given by

$$P_\Theta = \frac{1}{2\pi}\int_{-\infty}^{\infty} S_\Theta(\omega)d\omega = \frac{\Delta_{FM}^2}{2\pi}\int_{-\infty}^{\infty} \frac{S_M(\omega)}{\omega^2}d\omega. \qquad (5.16)$$

For the phase modulation case, recall that

$$\theta(t) = \Delta_{PM} \cdot m(t)dt,$$

therefore

$$S_\Theta(\omega) = \Delta_{PM}^2 S_M(\omega). \tag{5.17}$$

and the phase-modulated carrier power spectrum density is given by

$$\begin{aligned} S_S(\omega) &= \frac{\pi A^2 (1 - P_\Theta)}{2} [\delta(\omega + \omega_0) + \delta(\omega - \omega_0)] \\ &+ \frac{\Delta_{PM}^2 A^2}{4} [S_M(\omega + \omega_c) + S_M(\omega - \omega_c)] \end{aligned} \tag{5.18}$$

The spectrum of the phase-modulated signal is similar to the spectrum of the amplitude-modulated signal, and the bandwidth is the same, BW $= 2\omega_M$.

In this case, the power of the angle excitation signal is given by

$$P_\Theta = \frac{\Delta_{PM}^2}{2\pi} \int_{-\infty}^{\infty} S_M(\omega) d\omega. \tag{5.19}$$

5.2.3 Medium Modulation Index

A modulation index in the interval $0.1 \leq \beta \leq 10$ implies that it is possible to keep more terms from the Taylor expansion of Equation (5.9). It can be shown that, for a Gaussian signal, for example, all terms of the sine expansion vanish, because they are joint moments of odd order (Papoulis, 1983a), and therefore

$$\begin{aligned} R_S(\tau) &= \frac{A^2}{2} \cos(\omega_c \tau) \cdot \mathrm{E} \left[1 - \frac{(-\theta(t) + \theta(t + \tau))^2}{2} \right. \\ &+ \left. \frac{(-\theta(t) + \theta(t + \tau))^4}{4!} + \cdots \right] \end{aligned}$$

Considering that the bandwidth of the modulated signal does not exceed four times the bandwidth of the message signal $m(t)$, it is possible to neglect all terms of Expansion (5.20) of order four and higher.

Expanding the binomial in the first term, one obtains

$$\begin{aligned} R_S(\tau) &= \frac{A^2}{2} \cos(\omega_c \tau) \cdot \mathrm{E} \left[1 - \frac{\theta^2(t)}{2} - \frac{\theta^2(t + \tau)}{2} + \theta(t)\theta(t + \tau) \right] \\ &+ \frac{A^2}{2} \cos(\omega_c \tau) \cdot \mathrm{E} \left[\frac{\theta^4(t) + \theta^4(t + \tau)}{24} \right] \\ &+ \frac{A^2}{2} \cos(\omega_c \tau) \cdot \mathrm{E} \left[\frac{\theta^2(t)\theta^2(t + \tau)}{4} \right] \\ &- \frac{A^2}{2} \cos(\omega_c \tau) \cdot \mathrm{E} \left[\frac{\theta^2(t)\theta(t + \tau) + \theta(t)\theta^2(t + \tau)}{6} \right] + \dots \end{aligned} \tag{5.20}$$

It is observed that the contribution of the remaining terms of Equation (5.20) is less significant than the first ones, because of the rapid factorial increase in the denominator. Computing the expected values, assuming that the signal is stationary, gives

$$
\begin{aligned}
R_S(\tau) &= \frac{A^2}{2}\cos(\omega_c\tau)\cdot[1 - R_\Theta(0) + R_\Theta(\tau)] \\
&+ \frac{A^2}{24}\cos(\omega_c\tau)\cdot\mathrm{E}\left[\theta^4(t)\right] + \frac{A^2}{8}\cos(\omega_c\tau)\cdot\mathrm{E}\left[\theta^2(t)\theta^2(t+\tau)\right] \\
&- \frac{A^2}{12}\cos(\omega_c\tau)\cdot\mathrm{E}\left[\theta^2(t)\theta(t+\tau) + \theta(t)\theta^2(t+\tau)\right].
\end{aligned}
\tag{5.21}
$$

This is a generalized expression that includes the important terms for a medium modulation index. To compute the expected value of the higher order terms, it is necessary to assume that the modulating signal is Gaussian (McMahon, 1964).

For a Gaussian signal, the previous expression can be simplified, using Price's theorem, defined as (Price, 1958)

$$
\mathrm{E}[X^k Y^r] = kr\int_0^c \mathrm{E}[X^{k-1}Y^{r-1}]dc + \mathrm{E}[X^k]\mathrm{E}[Y^r]
\tag{5.22}
$$

in which

$$
c = \mathrm{E}[XY] - \mathrm{E}[X]\mathrm{E}[Y].
\tag{5.23}
$$

Applying Equations (5.22) and (5.23) in (5.21), it follows that

$$
\begin{aligned}
R_S(\tau) &= \frac{A^2}{2}\cos(\omega_c\tau)\cdot[1 - R_\Theta(0) + R_\Theta(\tau)] \\
&+ \frac{A^2}{4}\cos(\omega_c\tau)\cdot\left[P_\Theta^2 + R_\Theta^2(\tau) - 2P_\Theta R_\Theta(\tau)\right],
\end{aligned}
\tag{5.24}
$$

in which $P_\Theta = \mathrm{E}[\theta^2(t)]$.

Fourier transforming this expression, for the case of frequency modulation, gives

$$
\begin{aligned}
S_S(\omega) &= \frac{\pi A^2}{2}(1 - P_\Theta + \frac{P_\Theta^2}{2})[\delta(\omega + \omega_c) + \delta(\omega - \omega_c)] \\
&+ \frac{\Delta_{FM}^2 A^2}{4}(1 - P_\Theta)\left[\frac{S_M(\omega + \omega_c)}{(\omega + \omega_c)^2} + \frac{S_M(\omega - \omega_c)}{(\omega - \omega_c)^2}\right] \\
&+ \frac{\Delta_{FM}^4 A^2}{16\pi}\left[\frac{S_M(\omega + \omega_c)}{(\omega + \omega_c)^2} * \frac{S_M(\omega + \omega_c)}{(\omega + \omega_c)^2}\right. \\
&+ \left.\frac{S_M(\omega - \omega_c)}{(\omega - \omega_c)^2} * \frac{S_M(\omega - \omega_c)}{(\omega - \omega_c)^2}\right].
\end{aligned}
\tag{5.25}
$$

The convolution operations broaden the spectrum. The use of the previous formula permits to find the spectrum for any modulation index, just including more terms in Equation (5.20). The final bandwidth is obtained after the composition of several spectra. The power of the angle excitation signal is again given by Formula (5.16).

Figure 5.1 illustrates the FM spectrum for a sinusoidal message signal, considering a medium modulation index.

Fourier transforming Equation (5.24), for the phase modulation case, gives

$$
\begin{aligned}
S_S(\omega) &= \frac{\pi A^2}{2}(1 - P_\Theta + \frac{P_\Theta^2}{2})[\delta(\omega + \omega_c) + \delta(\omega - \omega_c)] \\
&+ \frac{\Delta_{PM}^2 A^2}{4}(1 - P_\Theta)\left[S_M(\omega + \omega_c) + S_M(\omega - \omega_c)\right] \\
&+ \frac{\Delta_{PM}^4 A^2}{16\pi}\left[S_M(\omega + \omega_c) * S_M(\omega + \omega_c)\right. \\
&+ \left. S_M(\omega - \omega_c) * S_M(\omega - \omega_c)\right].
\end{aligned}
\tag{5.26}
$$

The power of the angle excitation signal is given by Formula (5.19), as in the low modulation index case.

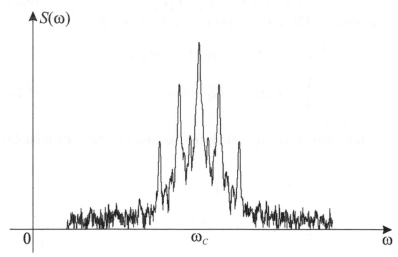

Figure 5.1   Spectrum of an FM signal for a sinusoidal message signal and medium modulation index.

5.2.4 High Modulation Index

For a high modulation index, $\beta > 10$, the power spectrum density of $s(t)$ assumes an unexpected shape, because it approaches the pdf of $m(t)$ in Equation (5.3), as the modulation index increases. That is the result of Woodward's theorem, as discussed in the following (Woodward, 1952).

A high modulating index also causes a spectrum broadening of the modulated signal, but this effect is limited to the extent of the deviation ratio (Blachman and McAlpine, 1969). The analog commercial FM and the analog wireless telephone can be considered as high modulation index systems (Lee, 1989).

For the high modulation index case, it is more interesting to use Euler's formula and rewrite Equation (5.8) as

$$R_S(\tau) = \frac{A^2}{4} e^{j\omega_c \tau} \mathrm{E}[e^{j(-\theta(t)+\theta(t+\tau))}]$$
$$+ \frac{A^2}{4} e^{-j\omega_c \tau} \mathrm{E}[e^{j(\theta(t)-\theta(t+\tau))}]. \tag{5.27}$$

In order to simplify Equation (5.27), the second-order linear mean square estimate of the process $\theta(t + \tau)$ is used.

As in the case of derivative and proportional stochastic control, the future estimate of the modulating signal $\theta(t + \tau)$ includes its current value $\theta(t)$ and its derivative $\theta'(t)$,

$$\theta(t + \tau) \approx \alpha\,\theta(t) + \beta\,\theta'(t). \tag{5.28}$$

Of course, there is an associated estimation error. The mean square error is given by

$$e(t, \alpha, \beta) = \mathrm{E}\left[\left(\theta(t+\tau) - \alpha\,\theta(t) - \beta\,\theta'(t)\right)^2\right]. \tag{5.29}$$

In order to obtain the values of the optimization parameters α and β, the derivative of the error must converge to zero. The partial derivative of the error in terms of α gives

$$\frac{\partial e(t, \alpha, \beta)}{\partial \alpha} = \mathrm{E}\left[\left(\theta(t+\tau) - \alpha\,\theta(t) - \beta\,\theta'(t)\right)\theta(t)\right] = 0, \tag{5.30}$$

which indicates that the minimum error is orthogonal to the random process. Recognizing the autocorrelations in the previous expression yields

$$R_\Theta(\tau) - \alpha R_\Theta(0) - \beta R_{\Theta'\Theta}(0) = 0,$$

The last term of the equation is zero, because the autocorrelation has a maximum at the origin. Thus,

$$\alpha = \frac{R_\Theta(\tau)}{R_\Theta(0)}. \tag{5.31}$$

The partial derivative of the error as a function of β gives

$$\frac{\partial e(t, \alpha, \beta)}{\partial \beta} = \mathrm{E}\left[\left(\theta(t + \tau) - \alpha\,\theta(t) - \beta\,\theta'(t)\right)\theta'(t)\right] = 0 \tag{5.32}$$

or

$$R_{\Theta'\Theta}(\tau) - \alpha R_{\Theta'\Theta}(0) - \beta R_{\Theta'}(\tau) = 0.$$

The first term of the expression is the derivative of the autocorrelation function in relation to τ. The second term is zero, as explained previously, and the third term is the negative of the second derivative of the autocorrelation function in relation to τ. Thus,

$$\beta = \frac{R'_\Theta(\tau)}{R''_\Theta(0)}. \tag{5.33}$$

Therefore, the best approximation to the future value of the modulating signal, in the mean square sense, is

$$\theta(t + \tau) \approx \frac{R_\Theta(\tau)}{R_\Theta(0)}\theta(t) + \frac{R'_\Theta(\tau)}{R''_\Theta(0)}\theta'(t). \tag{5.34}$$

Considering that the random process is slowly varying, compared to the spectral frequency of the modulated carrier, leads the approximation

$$R_\Theta(\tau) \approx R_\Theta(0).$$

Expanding the derivative of the autocorrelation in a Taylor series gives

$$R'_\Theta(\tau) \approx R'_\Theta(0) + \tau R''_\Theta(0). \tag{5.35}$$

Recalling that the autocorrelation has a maximum at the origin simplifies the approximation to

$$R'_\Theta(\tau) \approx \tau R''_\Theta(0).$$

Finally, substituting the previous approximations, the future value of the random process can be approximated by

$$\theta(t + \tau) \approx \theta(t) + \tau\,\theta'(t). \tag{5.36}$$

Use of the linear mean square estimator in Equation (5.27) then gives (Papoulis, 1983b)

$$R_S(\tau) = \frac{A^2}{4}e^{j\omega_c\tau}\mathrm{E}[e^{j\tau\theta'(t)}] + \frac{A^2}{4}e^{-j\omega_c\tau}\mathrm{E}[e^{-j\tau\theta'(t)}]. \qquad (5.37)$$

But, $\theta'(t) = d\theta(t)/dt = \omega(t)$, in which $\omega(t)$ is the carrier angular frequency deviation, at the instantaneous carrier angular frequency, thus (Alencar, 1989)

$$R_S(\tau) = \frac{A^2}{4}e^{j\omega_c\tau}\mathrm{E}[e^{j\tau\omega(t)}] + \frac{A^2}{4}e^{-j\omega_c\tau}\mathrm{E}[e^{-j\tau\omega(t)}]. \qquad (5.38)$$

Taking into account that

$$P_\Omega(\tau) = \mathrm{E}[e^{-j\tau\omega(t)}] = \int_{-\infty}^{\infty} p_\Omega(\omega(t))e^{-j\tau\omega(t)}\,d\omega(t) \qquad (5.39)$$

represents the characteristic function of process $\omega(t) = \theta'(t)$ and $p_\Omega(\omega(t))$ is its pdf, Equation (5.38) can be written as

$$R_S(\tau) = \frac{A^2}{4}e^{j\omega_c\tau}P_\Omega(-\tau) + \frac{A^2}{4}e^{-j\omega_c\tau}P_\Omega(\tau). \qquad (5.40)$$

If the pdf is considered symmetrical, without essential loss of generality, one obtains

$$R_S(\tau) = \frac{A^2}{2}P_\Omega(\tau)\cos(\omega_c\tau). \qquad (5.41)$$

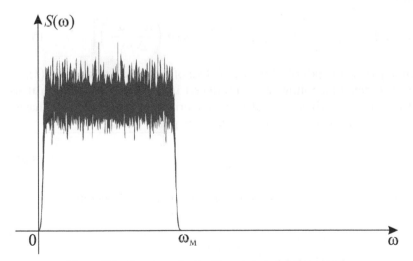

Figure 5.2 Spectrum for the Gaussian modulating signal.

Calculating the Fourier transform of Equation (5.41), it follows that

$$S_S(\omega) = \frac{\pi A^2}{2}[p_\Omega(\omega + \omega_c) + p_\Omega(\omega - \omega_c)]. \tag{5.42}$$

Considering the definition of $\omega(t)$ from Equation (5.3), it is noticed that $\omega(t) = \Delta_{FM} \cdot m(t)$, thus

$$p_\Omega(\omega(t)) = \frac{1}{\Delta_{FM}}p_M\left(\frac{m(t)}{\Delta_{FM}}\right) \tag{5.43}$$

in which $p_M(\cdot)$ is the pdf of $m(t)$, which is assumed stationary.

Substituting (5.43) into (5.42) gives, finally, the formula for the power spectrum density for the wideband frequency modulated signal (Alencar and Neto, 1991)

$$S_S(\omega) = \frac{\pi A^2}{2\Delta_{FM}}\left[p_M\left(\frac{\omega + \omega_c}{\Delta_{FM}}\right) + p_M\left(\frac{\omega - \omega_c}{\Delta_{FM}}\right)\right]. \tag{5.44}$$

Following a similar line of thought, it is possible to derive a formula for the spectrum of the phase-modulated signal (PM). It is instructive to recall that the instantaneous angular frequency is given by

$$\omega(t) = \frac{\theta(t)}{dt} = \Delta_{PM} \cdot m(t). \tag{5.45}$$

Therefore,

$$S_S(\omega) = \frac{\pi A^2}{2\Delta_{PM}}\left[p_{M'}\left(\frac{\omega + \omega_c}{\Delta_{PM}}\right) + p_{M'}\left(\frac{\omega - \omega_c}{\Delta_{PM}}\right)\right], \tag{5.46}$$

in which $p_{M'}(\cdot)$ is the pdf of of the derivative of the message signal $m(t)$.

The examples that follow were produced in a laboratory, using a thermal noise generator, whose output is characterized as a Gaussian stochastic process. For a Gaussian modulating signal, the pdf is given by

$$p_M(m) = \frac{1}{\sqrt{2\pi P_M}}e^{-\frac{m^2}{2P_M}}, \tag{5.47}$$

in which $P_M = R_M(0)$ denotes the power of signal $m(t)$, therefore

$$p_M(m) = \frac{1}{\sqrt{2\pi R_M(0)}}e^{-\frac{m^2}{2R_M(0)}}. \tag{5.48}$$

Figures 5.3–5.5, were obtained from a spectrum analyzer, varying the modulating index at the input by changing the modulating signal power. They illustrate the changes undergone by the modulated carrier spectrum as the modulation index is increased, for a Gaussian modulating signal.

The power spectrum density of the modulating signal, which can be considered a white Gaussian noise, is shown in Figure 5.2. The spectrum is approximately flat, and the signal has bandwidth $BW = \omega_M$.

From Equation (5.50), and considering the property

$$R_{M'}(\tau) = -R_M''(\tau), \tag{5.49}$$

that is, the autocorrelation of the signal derivative equals the negative of the second derivative of the autocorrelation function, and the fact that the response of a linear system to a Gaussian input is a Gaussian signal, one can determine the pdf for the derivative of a Gaussian signal, which is given by

$$p_{M'}(m) = \frac{1}{\sqrt{-2\pi R_M''(0)}} e^{-\frac{m^2}{(-2R_M''(0))}}. \tag{5.50}$$

That is, the spectrum of a wideband PM signal, when modulated by a Gaussian signal, presents similar characteristics to its FM counterpart.

For a narrowband modulation, the PM spectrum approximates the spectrum of the modulating signal, which is quite different from the FM spectrum for the same conditions.

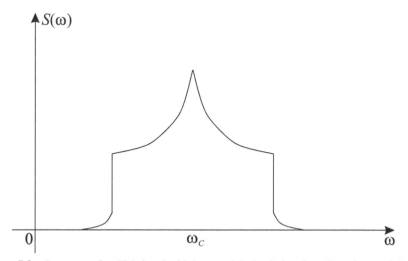

Figure 5.3   Spectrum of an FM signal with low modulation index, for a Gaussian modulating signal.

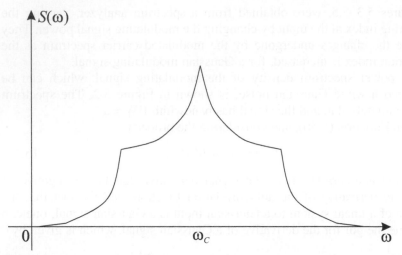

Figure 5.4   Spectrum of an FM signal for a Gaussian modulating signal and a medium modulation index.

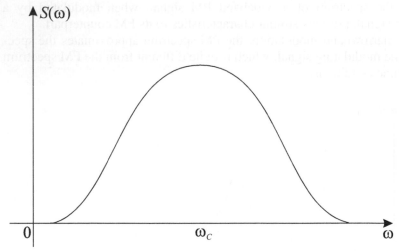

Figure 5.5   Spectrum of an FM signal for a Gaussian modulating signal and high modulation index.

For a high modulation index, the frequency deviation is given by $\Delta_{FM}\sqrt{P_M}$ and the bandwidth is approximated by the formula BW = $2\Delta_{FM}\sqrt{P_M}$, in order to include most of the modulated carrier power.

As previously derived, the bandwidth for a narrowband FM is $\text{BW} = 2\omega_M$. Interpolating between both values gives a formula that covers the whole range of modulation indices β

$$\text{BW} = 2\omega_M + 2\Delta_{FM}\sqrt{P_M} = 2\left(\frac{\Delta_{FM}\sqrt{P_M}}{\omega_M} + 1\right)\omega_M = 2(\beta + 1)\omega_M. \tag{5.51}$$

This is the well-known Carson's rule, whose heuristic deduction first appeared, in 1922, in a paper published by John Carson (Carson, 1922).

A sinusoidal signal $m(t) = V\sin(\omega_M t + \varphi)$, in which φ is uniformly distributed in the interval $(0, 2\pi]$, has the following pdf

$$p_M(m) = \frac{1}{\pi\sqrt{V^2 - m^2}}, |m| < V. \tag{5.52}$$

Consequently, a carrier modulated by a sinusoidal signal, with high modulation index, has the following spectrum

$$S_S(\omega) = \frac{1}{2\sqrt{(V\Delta_{FM})^2 - (\omega + \omega_c)^2}} + \frac{1}{2\sqrt{(V\Delta_{FM})^2 - (\omega - \omega_c)^2}}, \tag{5.53}$$

for $|w - \omega_c| < \Delta_{FM}V$.

The modulated signal occupies in this case, a bandwidth equivalent to $\text{BW} = 2\Delta_{FM}V$, and its PSD is shown in Figure 5.6.

For the phase-modulated signal, using the same sinusoid as before as the modulating signal, the derivative of the message signal is given by

$$m'(t) = \omega_M V\cos(\omega_M t + \varphi),$$

which has the following pdf

$$p_{M'}(m) = \frac{1}{\pi\sqrt{(\omega_M V)^2 - m^2}}, |m| < \omega_M V. \tag{5.54}$$

The spectrum can then be computed, resulting in

$$S_S(\omega) = \frac{1}{2\sqrt{(\omega_M V\Delta_{PM})^2 - (\omega + \omega_c)^2}}$$

$$+ \frac{1}{2\sqrt{(\omega_M V\Delta_{PM})^2 - (\omega - \omega_c)^2}}, \tag{5.55}$$

for $|w - \omega_c| < \omega_M\Delta_{PM}V$.

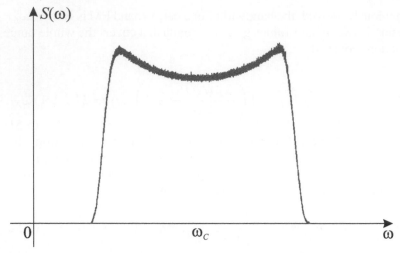

Figure 5.6 Spectrum of an FM signal for a sinusoidal modulating signal and high modulation index.

The bandwidth for the phase-modulated signal is BW$= 2\omega_M \Delta_{PM} V$ and its PSD is similar to the one shown in Figure 5.6.

The voice signal is modeled as a Gamma distribution (Paez and Glisson, 1972),

$$p_M(m) = \frac{\sqrt{k}}{2\sqrt{\pi}} \frac{e^{-k|m|}}{\sqrt{|m|}} \tag{5.56}$$

Therefore, the corresponding FM spectrum is given by

$$R_S(\omega) = \frac{A^2\sqrt{\pi k}}{4\sqrt{\Delta_{FM}}} \frac{e^{-\frac{k}{\Delta_{FM}}|w \pm \omega_c|}}{\sqrt{|w \pm \omega_c|}} \tag{5.57}$$

and its bandwidth is $\frac{2\Delta_{FM}\sqrt{0.75}}{k}$.

5.3 Frequency and Phase Demodulation

In order to recover the original signal, the modulated carrier must be demodulated, as shown in Figure 5.7. For the FM case, the incoming signal

$$s(t) = A\cos\left(\omega_c t + \theta(t) = \Delta_{FM} \cdot \int_{-\infty}^{t} m(t)dt + \phi\right) \tag{5.58}$$

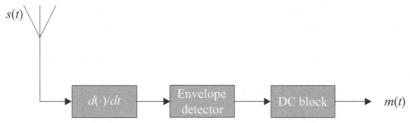

Figure 5.7   Block diagram of an FM demodulator.

is differentiated (passes through a discriminator), to give

$$s(t) = -A\left[\omega_c + \Delta_{FM} \cdot m(t)\right]\sin\left(\omega_c t + \theta(t) = \Delta_{FM} \cdot \int_{-\infty}^{t} m(t)dt + \phi\right).$$
(5.59)

The signal is envelope demodulated (non-coherent demodulation) and low-pass filtered, which results in

$$r(t) = -A\left[\omega_c + \Delta_{FM} \cdot m(t)\right]$$
(5.60)

The DC level then is blocked, to give the original signal $m(t)$ multiplied by a constant. A phase tracking loop, also called Phase Locked Loop (PLLs), is commonly used to demodulate FM signals.

Figure 5.8 illustrates the block diagram of PLL, which is composed of a mixer, followed by a low-pass filter with impulse response $f(t)$ and an output filter, whose impulse response is $g(t)$. The feedback loop has a voltage controlled oscillator (VCO), which produces an output $x(t)$ whose frequency is proportional to the amplitude of $z(t)$.

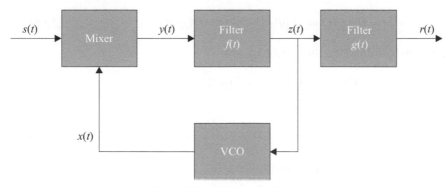

Figure 5.8   Phase locked loop.

The following analysis is performed to obtain the frequency response, or transfer function, of the PLL, in order to verify that it can actually demodulate the input signal.

The signal at the input of the PLL $s(t)$ represents the modulated carrier plus noise

$$s(t) = A \cos \left[w_c t + \theta(t) + \phi \right] + n(t). \tag{5.61}$$

The signal at the output of the VCO is

$$x(t) = \sin \left[w_c t + \gamma(t) + \phi \right]. \tag{5.62}$$

in which $\gamma(t)$ is proportional to the integral of $z(t)$,

$$\gamma(t) = \Delta_{VCO} \int_{-\infty}^{t} z(t) dt, \tag{5.63}$$

which implies that the VCO instantaneous frequency $w(t)$ is proportional to the signal amplitude

$$w(t) = \frac{d\gamma(t)}{dt} = \Delta_{VCO} z(t). \tag{5.64}$$

At the mixer output, the signal is

$$y(t) = A \sin \left[\theta(t) - \gamma(t) \right] + n_M(t). \tag{5.65}$$

in which $n_M(t)$ represents the noise filtered by the mixer.

The difference between the modulated carrier phase and the VCO signal phase $\theta(t) - \gamma(t)$ must be kept to a minimum, for the PLL to work properly. Therefore, at the mixer output, the signal can be approximated by

$$y(t) = A \left[\theta(t) - \gamma(t) \right] + n_M(t). \tag{5.66}$$

Fourier transforming Equation (5.66) gives

$$Y(\omega) = A \left[\Theta(\omega) - \Gamma(\omega) \right] + N_M(\omega). \tag{5.67}$$

in which

$$\Gamma(\omega) = \Delta_{VCO} \frac{Z(\omega)}{j\omega} \tag{5.68}$$

and $N_M(\omega)$ is the Fourier transform of $n_M(t)$.

At the output of the low-pass filter,

$$Z(\omega) = AF(\omega)\left[\Theta(\omega) - \Gamma(\omega)\right] + F(\omega)N_M(\omega) \qquad (5.69)$$

in which $F(\omega)$ is the transfer function of the filter.

Substituting the Fourier transform of the VCO phase yields

$$Z(\omega) = AF(\omega)\left[\Theta(\omega) - \Delta_{VCO}\frac{Z(\omega)}{j\omega}\right] + F(\omega)N_M(\omega) \qquad (5.70)$$

Solving for $Z(\omega)$ gives

$$Z(\omega) = \left[\frac{j\omega AF(\omega)}{A\Delta_{VCO}F(\omega) + j\omega}\right]\left[\Theta(\omega) + \frac{N_M(\omega)}{A}\right]. \qquad (5.71)$$

The Fourier transform of the signal at the output of the PLL is

$$R(\omega) = G(\omega)Z(\omega). \qquad (5.72)$$

For an FM signal,

$$\Theta(\omega) = \Delta_{FM}\frac{M(\omega)}{j\omega} \qquad (5.73)$$

in which $M\omega)$ is the Fourier transform of the modulating signal $m(t)$.

Then, the transfer function of the output filter, to correctly demodulate the signal, is

$$G(\omega) = \frac{A\Delta_{VCO}F(\omega) + j\omega}{AF(\omega)}, \qquad (5.74)$$

which is valid for $|\omega| \leq \omega_M$, and null outside this interval, in which ω_M is the maximum frequency of the modulating signal.

This results in

$$R(\omega) = \Delta_{FM}M(\omega) + \frac{j\omega N_M(\omega)}{A}. \qquad (5.75)$$

Thus, the transfer function of the PLL for the signal is just Δ_{FM}. For the noise, the transfer function is $\frac{j\omega}{A}$. Recalling that, at the output of a linear filter, the PSD is given by

$$S_Y(\omega) = |H(\omega)|^2 S_X(\omega)$$

and that the input noise has a uniform PSD N_0, which implies that the noise at the output of the PLL has a quadratic PSD.

$$S_N(\omega) = \frac{\omega^2 N_0}{A^2}. \qquad (5.76)$$

For a PM signal,

$$G(\omega) = \frac{A\Delta_{VCO}F(\omega) + j\omega}{AF(\omega)/j\omega}, \tag{5.77}$$

which is valid in the interval $|\omega| \leq \omega_M$, and null outside this interval.

The output signal transfer function is

$$R(\omega) = \Delta_{PM}M(\omega) + \frac{N_M(\omega)}{A}. \tag{5.78}$$

The noise PSD for the PM demodulator is then

$$S_N(\omega) = \frac{N_0}{A^2}, \tag{5.79}$$

which represents white noise.

5.4 Performance Evaluation of Angle Modulation

The preceding analysis has shown a bandwidth equivalent to $BW = 2\omega_M$, for narrowband angle modulation and a modulating signal of bandwidth ω_M.

For wideband frequency modulation, considering the upper limit for the modulation index, the bandwidth is $BW = 2\sigma_F = 2\Delta_{FM}\sigma_M$, in which σ_F represents the RMS frequency deviation from the carrier spectral frequency and σ_M is the RMS value of the Gaussian modulating signal.

The modulation index, for a random modulating signal, has been defined as

$$\beta = \frac{\sigma_F}{\omega_M} = \frac{\Delta_{FM}\sigma_M}{\omega_M} = \frac{\Delta_{FM}\sqrt{P_M}}{\omega_M}, \tag{5.80}$$

and recall that Δ_{FM} is the frequency deviation index and P_M is the message signal average power.

The approximate formula for the bandwidth, which can be used for both narrowband and wideband case, is

$$BW = 2\omega_M + 2\Delta_{FM}\sigma_M = 2\omega_M\left(1 + \frac{\Delta_{FM}\sigma_M}{\omega_M}\right) = 2(\beta+1)\omega_M. \tag{5.81}$$

As mentioned earlier, Formula (5.81) is known as Carson's rule to determine the bandwidth of an FM signal. This formula is used to compute the demodulator, or front-end filter, bandwidth, as well as the intermediate filter (IF) bandwidth.

The received signal power is $P_S = \frac{A^2}{2}$. Noise is also received by the demodulator, and it is considered to have a uniform power spectrum density

N_0. The noise is filtered by the front-end filter, which gives a net received power of

$$P_N = \frac{N_0(\beta + 1)\omega_M}{\pi}. \tag{5.82}$$

The signal to noise ratio at the input of the demodulator is given by

$$\text{SNR}_I = \frac{A^2/2}{N_0(\beta + 1)\omega_M/\pi} = \frac{\pi A^2}{2N_0(\beta + 1)\omega_M}. \tag{5.83}$$

The demodulator consists of a discriminator, which is equivalent to a circuit to compute the derivative of the input signal, operating at the intermediate frequency, and produces at its output a signal whose power is $P_{\hat{M}} = \Delta_{FM}^2 P_M$.

Noise is, as well, differentiated by the discriminator, and its output power spectrum density is given by the formula $S_{N'}(\omega) = |H(\omega)|^2 S_N(\omega)$, in which $H(\omega) = \frac{j\omega}{A}$ represents the discriminator, or the PLL, transfer function and $S_N(\omega) = N_0$.

At the output of the demodulator, after being filtered by a bandpass filter of bandwidth ω_M, the noise power is given by

$$P_{N'} = \frac{1}{2\pi} \int_{-\omega_M}^{\omega_M} S_{N'}(\omega)d\omega = \frac{1}{2\pi} \int_{-\omega_M}^{\omega_M} |H(\omega|^2 S_N(\omega)d\omega. \tag{5.84}$$

Substituting the previous results into (5.84), it follows that

$$P_{N'} = \frac{1}{2\pi} \int_{-\omega_M}^{\omega_M} \frac{N_0\omega^2}{A^2} d\omega = \frac{N_0\omega^3}{3\pi A^2}. \tag{5.85}$$

The signal to noise ratio at the demodulator output is then

$$\text{SNR}_O = \frac{3\pi A^2 \Delta_{FM}^2 P_M}{N_0\omega_M^3}. \tag{5.86}$$

Finally, the demodulation gain for the frequency demodulator is given by

$$\eta = \frac{\text{SNR}_O}{\text{SNR}_I} = 6\beta^2(\beta + 1). \tag{5.87}$$

A similar derivation for the PM demodulation scheme, leads to

$$\eta = \frac{\text{SNR}_O}{\text{SNR}_I} = 2\alpha^2(\alpha + 1), \tag{5.88}$$

in which $\alpha = \Delta_{PM}\sqrt{P_M}$ represents the PM modulation index.

Formulas (5.87) and (5.88) show that the demodulation gain increases with the square of the modulation index, for narrowband FM and PM, and with the third power of the index, for wideband FM and PM.

This demonstrates the importance of the modulation index to improve the quality of the angle-modulated signal reception. It is worthy to mention that the bandwidth increases in the same proportion as the modulation index, which imposes a tradeoff on the design of FM and PM transmitters. In order to improve reception, the bandwidth has to be increased, but the telecommunication regulators limit the amount of bandwidth allowed to each operator.

Phase modulation presents a similar result for the demodulation gain, as well as, the same considerations regarding the tradeoff between power and bandwidth. But, wideband PM is not so popular as wideband FM, because it requires absolute phase deviation measurement, instead of detection modulo 2π, and this limits its usefulness for practical purposes (Gagliardi, 1988).

5.5 Angle Modulation with a Digital Signal

Phase Shift Keying (PSK) is a special case of angle modulation with a digital signal. It can be described by the following equations:

$$s(t) = A\cos(\omega_c t + \Delta_{PM} m(t) + \phi) \tag{5.89}$$

$$m(t) = \sum_{j=-\infty}^{\infty} m_j p(t - jT_b), \tag{5.90}$$

in which m_j represent the random values of the modulating signal, T_b is the pulse period, and $p(t)$ is a unit pulse waveform. The power of the modulated carrier is $P_S = A^2/2$.

The constellation diagram of a PSK modulation scheme, with parameters $\Delta_{PM} = \pi/4$, $\phi = 0$, and $m_j \in \{0,1,2,3,4,5,6,7\}$, is illustrated in Figura 5.9.

The main feature of the PSK modulation is its carrier constant envelope, with the constellation symbols positioned on a circle. This gives a certain advantage for the scheme regarding non-linear amplifiers and multiplicative noise fading channels.

The transponder of a satellite has to operate in the region of maximum power output, that is, the non-linear part of the amplifier curve, because of the limitation in energy supply in space. This is the main reason to the success of the PSK scheme in the satellite business.

On the other hand, mobile communication systems, such as those owned by digital cellular operators, can be characterized as multiplicative noise

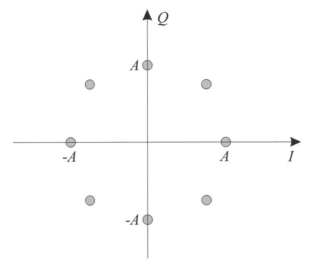

Figure 5.9 Constellation for a PSK signal.

fading channels, which usually require modulation schemes that are less sensitive to multiplicative noise.

An increase in the transmission rate, which implies the addition of new symbols, makes the PSK scheme more susceptible to noise, increasing the probability of error, because the symbols get closer in signal space. This modulation technique is not as efficient as the quadrature modulation to allocate symbols on the constellation diagram.

The probability of error for the coherent binary PSK is (Haykin, 1988)

$$P_e = \frac{1}{2}\text{erfc}\left(\sqrt{\frac{E_b}{N_0}}\right), \tag{5.91}$$

in which E_b is the pulse energy and N_0 represents the PSD.

For the coherent FSK, the probability of error has a 3 dB penalty, compared to the PSK result (Haykin, 1988)

$$P_e = \frac{1}{2}\text{crfc}\left(\sqrt{\frac{E_b}{2N_0}}\right). \tag{5.92}$$

The noncoherent FSK has the probability of error (Haykin, 1988)

$$P_e = \frac{1}{2}\exp\left(\frac{E_b}{2N_0}\right). \tag{5.93}$$

6

Digital Modulation Theory

6.1 Introduction

This chapter presents the general principles used to analyze digital communication schemes. Digital modulation is a mapping of symbols into waveforms, to transmit information in a channel.

Modulation schemes with varying carrier envelope, such as QAM and ASK, are subject to amplitude variations, because of non-linear amplification, fading, and interference. Angle modulation occupies a large bandwidth, if resilience against noise is required. On the other hand, large constellations, with large transmission rates, are more susceptible to noise and fading, while small ones are more robust, but support only low bit rates.

6.2 Signal Space

Digital modulation codes a finite bit sequence into a waveform, from a certain set of signals. The receiver minimizes the error probability if it decodes a received signal that is near the transmitted one, under a given distance metric.

The usual procedure to represent the signal is to project it in a set of basis functions, to obtain a vectorial representation of each waveform. This restricts the dimensionality of the signal space and permits the use of regular vectorial distance metrics.

6.2.1 System Model

Consider the communication system shown in Figure 6.1, which is subject to AWGN, that is stationary, zero mean, and has PSD $S_N(\omega) = N_0/2$.

The channel impulse response is $h(t) = \delta(t)$ and very T seconds the source selects a message m_i from a set $\mathcal{M} = \{m_1, \cdots, m_M\}$ to be transmitted. Each message is chosen with probability p_i, such that $\sum_i p_i = 1$.

Because the message set $\mathcal{M}$ has cardinality M, each transmitted message carries, at most, $R = \log_2 M/T$ bits of information per second. The $\log_2 M$

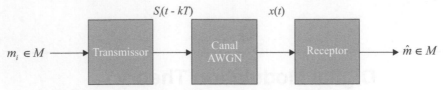

Figure 6.1   Communication system model.

bits corresponding to message m_i are coded as a signal $s_i(t) \in \mathcal{S} = \{s_1(t), \cdots, s_M(t)\}$ with energy

$$E_i = \int_T s_i^2(t)dt, \quad i = 1, \cdots, M. \tag{6.1}$$

The transmitted signal is, therefore, given by

$$\sum_k s_i(t - kT), \tag{6.2}$$

in which $s_i(t)$ is an analog message that represents message m_i in the transmission interval $[kT, (k+1)T]$. The received signal, corresponding to the original message m_i, in the interval $[kT, (k+1)T]$ is $x(t) = s_i(t-kT)+n(t)$. For each transmitted signal, $s_i(t-kT)$, the receiver must determine the best estimate of $s_i(t) \in \mathcal{S}$, which is equivalent to minimize the error probability for each symbol

$$P_e = \sum_{i=1}^{M} p_i P(\hat{m} \neq m_i | m_i \text{ sent}), \tag{6.3}$$

in each time interval $[kT, (k+1)T]$.

A geometrical representation of the set of signal is a manner to solve the problem of finding the optimum receiver, subject to AWGN, based on a minimum distance criterion.

6.2.2 Representation by Basis Functions

The representation of signals by basis functions uses the Gram–Schmidt orthogonalization procedure (Wozencraft and Jacobs, 1965b), which states that it is possible to represent a set of M waveforms $\mathcal{S} = (s_1(t), \cdots, s_M(t))$ defined in the interval $[0, T]$ by a linear combination of $N \leq M$ orthonormal basis functions $\{\phi_1(t), \cdots, \phi_N(t)\}$.

Therefore, it is possible to write each waveform $s_i(t) \in \mathcal{S}$, as

$$s_i(t) = \sum_{i=1}^{N} s_{ij}\phi_j(t), \ 0 \le t < T, \tag{6.4}$$

in which

$$s_{ij} = \int_0^T s_i(t)\phi_j(t)dt \tag{6.5}$$

is a real coefficient that represents the projection of $s_i(t)$ over the basis function $\phi_j(t)$, and

$$\int_0^T \phi_i(t)\phi_j(t)dt = \left\{ \begin{array}{l} 1, i = j \\ 0, i \ne j. \end{array} \right. \tag{6.6}$$

If the functions $\{s_i(t)\}$ are linearly independent, then $N = M$; otherwise, $N < M$. For most modulation techniques, the basis set has only two functions ($N = 2$) that correspond to the in-phase and quadrature dimensions of the transmitted signal.

Coefficients $\{s_{ij}\}$ are denoted as $s_i = (s_{i1}, \cdots, s_{iN})$, in the vector space. Given the basis functions, $\{\phi_1(t), \cdots, \phi_N(t)\}$, there is one to one correspondence between the signal $s_i(t)$ and its vectorial representation s_i.

Vector s_i defines the signal constellation corresponding to message m_i. As discussed previously, signal constellations for the usual modulation schemes, such as M-PSK and M-QAM, are bi-dimensional. The noise signal is also projected into the signal space, and this simplifies the performance analysis of the system as well as the optimum receiver design.

Before converting the model described in Figure 6.1 in a vectorial model, it is convenient to present some definitions to characterize a vector in the signal space $\mathbb{R}^N$.

The length of a vector in the signal space is defined as

$$||s_i||^2 = s_i^T s_i = \sum_{k=1}^{N} s_{ik}^2. \tag{6.7}$$

The square root of this length, $||s_i||$, is the vector norm.

The distance between two vectors s_i and s_j is

$$||s_i - s_j||^2 = \sum_{k=1}^{N}(s_{ik} - s_{jk})^2 = \int_0^T (s_i(t) - s_j(t))^2 dt, \tag{6.8}$$

in which the second equality is obtained by writing $s_i(t)$ and $s_j(t)$ as basis functions in Equation (6.4), and using the orthonormality property of the basis functions.

The inner product, $< s_i(t), s_j(t) >$, between two real signals $s_i(t)$ and $s_j(t)$ in the interval $[0, T]$ is defined as

$$< s_i(t), s_j(t) > = \int_0^T s_i(t)s_j(t)dt. \tag{6.9}$$

In a similar manner, the inner product $< \boldsymbol{s}_i, \boldsymbol{s}_j >$ between two real vectors is

$$< \boldsymbol{s}_i, \boldsymbol{s}_j > = \boldsymbol{s}_i^T \boldsymbol{s}_j = \int_0^T s_i(t)s_j(t)dt = < s_i(t), s_j(t) > . \tag{6.10}$$

Two signals are considered orthogonal, if their inner product is zero.

6.2.3 Receiver Design and Sufficient Statistic

Given the output of a channel $x(t) = s_i(t) + n(t), 0 \le t < T$, one can design a receiver to infer which m_i (or $s_i(t)$) has been sent in the time interval $[0, T]$, and a similar procedure is attempted for each time interval $[kT, (k+1)T]$.

Consider the receiver structure shown in Figure 6.2, in which

$$s_{ij} = \int_0^T s_i(t)\phi_j(t)dt \tag{6.11}$$

and

$$n_j = \int_0^T n(t)\phi_j(t)dt. \tag{6.12}$$

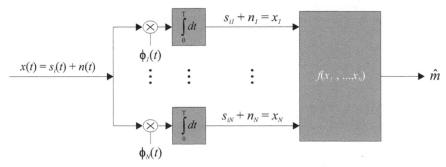

Figure 6.2   Receiver structure to detect a signal subject to AWGN.

It is possible to write $x(t)$ as

$$\sum_{j=1}^{N}(s_{ij} + n_j)\phi_j(t) + n_r(t) = \sum_{j=1}^{N} x_j\phi_j(t) + n_r(t), \qquad (6.13)$$

in which $x_j = s_{ij} + n_j$ and $n_r(t) = n(t) - \sum_{j=1}^{N} n_j\phi_j(t)$ denote the remaining noise.

The receiver obtains an estimate $\hat{m}$ of the transmitted message m_i as a function of $x_1, \cdots, x_N$. Therefore, the receiver ignores any information contained in the remaining noise term $n_r(t)$, which is the noise component orthogonal to the signal space generated by the basis set $(\phi_1(t), \cdots, \phi_N(t))$.

If it is possible to prove that $n_r(t)$ is useless to decide if $s_i(t)$ has been sent, then it is only necessary to determine the optimum receiver to $(x_1, \cdots, x_N)$, once it contains all information needed to the optimum detection of $s_i(t)$ given $x(t)$. In other words, $(x_1, \cdots, x_N)$ constitutes a sufficient statistic to detect $s_i(t)$.

Because $n(t)$ is a Gaussian random process, the channel output $x(t) = s_i(t) + n(t)$ is also a Gaussian random process. and the random vector $(x_1, \cdots, x_N)$ is also Gaussian.

Recalling that $x_j = s_{ij} + n_j$, the mean of x_j is given by

$$\mu_{x_j} = E[x_j] = E[s_{ij} + n_j] = s_{ij}, \qquad (6.14)$$

and because $n(t)$ has zero mean, the variance is

$$\sigma_{x_j} = E[x_j - \mu_{x_j}]^2 = E[s_{ij} + n_j - s_{ij}]^2 = E[n_j^2]. \qquad (6.15)$$

Therefore, one obtains,

$$\begin{aligned}
\mathrm{Cov}[x_j x_k] &= E[(x_j - \mu_{x_j})(x_k - \mu_{x_k})] = E[n_j n_k] \\
&= E\left[\int_0^T n(t)\phi_j(t)dt \int_0^T n(\tau)\phi_k(\tau)d\tau\right] \\
&= \int_0^T \int_0^T E[n(t)n(\tau)]\phi_j(t)\phi_k(\tau)dtd\tau \\
&= \int_0^T \int_0^T \frac{N_0}{2}\delta(t-\tau)\phi_j(t)\phi_k(\tau)dtd\tau = \frac{N_0}{2}\int_0^T \phi_j(t)\phi_k(\tau)dt \\
&= \begin{cases} N_0/2, i = j \\ 0, i \neq j \end{cases}
\end{aligned}$$

$$(6.16)$$

in which the last equality is a result of the basis functions orthogonality. The terms x_i are uncorrelated, and independent, because they are Gaussian (Papoulis, 1984). Furthermore, $E[n_j^2] = N_0/2$.

If the random vector corresponding to the correlator output is defined as $x = [x_1, \cdots, x_N]$, given that message m_i is transmitted, then x_i has Gaussian distribution, with average s_{ij} and variance $N_0/2$.

Therefore, by the independence of the terms x_i,

$$p(\boldsymbol{x}|m_i) = \prod_{j=1}^{N} p(x_j|m_i) = \frac{1}{(\pi N_0)^{N/2}} \exp\left[-\frac{1}{N_0} \sum_{j=1}^{N} (x_j - s_{ij})^2\right].$$

(6.17)

It is possible to show that if $E[x_j n_r(t_k)] = 0$ for any $t_k \colon 0 \le t_k < T$. Because every random process is completely characterized by its set of time samples, then x_j is independent of any function of the remaining noisy process $n_r(t)$. Besides, as the transmitted signal is independent of the channel noise, s_{ij} is independent of $n_r(t)$.

The main goal of the receiver design is to minimize the transmitted message, m_i, detection error probability given the received signal $x(t)$. To minimize $P_e = p(\hat{m} \ne m_i | x(t)) = 1 - p(\hat{m} = m_i | x(t))$, one needs to maximize $p(\hat{m} = m_i | x(t))$.

Therefore, the receiver output, $\hat{m}$, given the received signal $x(t)$ must correspond to the message m_i that maximizes $p(m_i \text{ sent}|x(t))$. Recall that $x(t)$ is completely described by $\boldsymbol{x} = (x_1, \cdots, x_N)$ and $n_r(t)$, and that the decision on which message has been transmitted must be a function of the remaining noisy process $f[n_r(t)]$, then

$$
\begin{aligned}
p(m_i \text{ sent}|x(t)) &= p((s_{i1}, \cdots, s_{iN}) \text{ sent}|(x_1, \cdots, x_N, f[n_r(t)])) \\
&= \frac{p((s_{i1}, \cdots, s_{iN}) \text{ sent}, (x_1, \cdots, x_N), f[n_r(t)]}{p((x_1, \cdots, x_N), f[n_r(t)])} \\
&= \frac{p((s_{i1}, \cdots, s_{iN}) \text{ sent}, (x_1, \cdots, x_N))p(f[n_r(t)])}{p((x_1, \cdots, x_N))p(f[n_r(t)])} \\
&= p((s_{i1}, \cdots, s_{iN}) \text{ sent}|(x_1, \cdots, x_N)),
\end{aligned}
$$

(6.18)

in which the third equality results from the fact that any function of the remaining noisy process is independent of $(x_1, \cdots, x_N)$ e $(s_{i1}, \cdots, s_{iN})$. Analysis shows that $(x_1, \cdots, x_N)$ is a sufficient statistic for $x(t)$, in the detection of m_i, because the error detection probability is minimized.

6.2.4 Maximum Likelihood Decision

In the previous section, it has been seen that the detection probability error is minimized if the detector output $\hat{m}$ is selected to maximize $1 - P_e = p(\hat{m} \text{ sent}|\boldsymbol{x})$. Using Bayes rule,

$$p(\hat{m} \text{ sent}|\boldsymbol{x}) = \frac{p(\boldsymbol{x}|\hat{m} \text{ sent})p(\hat{m} \text{ sent})}{p(\boldsymbol{x})}. \quad (6.19)$$

Assuming equiprobable messages ($p(m_i) = 1/M$), $\hat{m}$ is chosen as the argument that maximizes $p(\boldsymbol{x}|\hat{m})$:

$$\hat{m} = \arg\max_{m_i} p(\boldsymbol{x}|m_i), i = 1, \cdots, M. \quad (6.20)$$

The maximum likelihood function, associated to the receiver, is defined as

$$L(m_i) = p(\boldsymbol{x}|m_i \text{ sent}). \quad (6.21)$$

For a maximum likelihood receiver, $\hat{m}$ is chosen as the argument that maximizes $L(m_i)$. Because the logarithmic function is increasing, the maximization of $L(m_i)$ is equivalent to maximize the logarithm of the likelihood function, defined as

$$l(m_i) = \log L(m_i) = -\frac{1}{N_0} \sum_{j=1}^{N} (x_j - s_{ij})^2, \quad (6.22)$$

in which the second equality is obtained by the substitution of $L(m_i)$ into Equation (6.17).

Note that the log-likelihood function depends only on the distance between the received vector $\boldsymbol{x}$ and the set of points of the transmitted signal constellation, s_i, $i = 1, \cdots, M$.

Therefore, the maximum likelihood receiver computes $\boldsymbol{x}$ from $x(t)$ using the structure presented in Figure 6.2 and then decodes

$$\hat{m} = \arg\max_{m_i} l(m_i). \quad (6.23)$$

In a similar way, it is possible to find the signal constellation corresponding to the maximization of the message signal m_i in Equation (6.23), as

$$\arg\max_{s_i} \left(-\frac{1}{N_0} \sum_{j=1}^{N} (x_j - s_{ij})^2 \right) = \arg\max_{s_i} \left(-\frac{1}{N_0} ||\boldsymbol{x} - s_i||^2 \right). \quad (6.24)$$

The maximum likelihood receiver decodes the message m_i corresponding to the signal constellation, s_i, $i = 1, \cdots, M$, as the one which is nearest to the received vector x. It is possible to divide the signal space into M decision regions

$$Z_i = (x: \|x - s_i\| < \|x - s_j\|, \forall j = 1, \cdots, M, j \neq i) \quad i = 1, \cdots, M. \tag{6.25}$$

Then, the maximum likelihood (ML) decoding only implies the determination of the decision region Z_i in which the received vector x is located, that is, $x \in Z_i \to \hat{m} = m_i$. Therefore, the function $f(x) = f(x_1, \cdots, x_N)$ in Figure 6.2 is given by $\hat{m} = m_i$: $x \in Z_i$.

It can be shown that Figure 6.2 is equivalent to the matched filter of Figure 6.3.

6.2.5 Error Probability and the Union Bound

Consider the error probability associated to the ML receiver structure. For equiprobable messages, $p(m_i \text{ sent}) = 1/M$, one has

$$P_e = \sum_{i=1}^{M} p(x \notin Z_i | m_i \text{ sent}) p(m_i \text{ sent})$$

$$= \frac{1}{M} \sum_{i=1}^{M} p(x \notin Z_i | m_i \text{ sent})$$

$$= 1 - \frac{1}{M} \sum_{i=1}^{M} p(x \in Z_i | m_i \text{ sent})$$

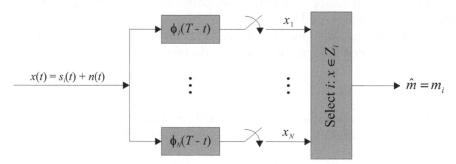

Figure 6.3   Matched filter structure.

$$= 1 - \frac{1}{M} \sum_{i=1}^{M} \int_{Z_i} p(\boldsymbol{x}|m_i)d\boldsymbol{x} \qquad (6.26)$$

$$= 1 - \frac{1}{M} \sum_{i=1}^{M} \int_{Z_i} p(\boldsymbol{x} = \boldsymbol{s}_i + \boldsymbol{n}|\boldsymbol{s}_i)d\boldsymbol{n}$$

$$= 1 - \frac{1}{M} \sum_{i=1}^{M} \int_{Z_i - \boldsymbol{s}_i} p(\boldsymbol{n})d\boldsymbol{n}.$$

The integrals in Expression (6.26) are defined in the subset N-dimensional $Z_i \subset \mathbb{R}^N$. The computation of the error probability is illustrated in Figure 6.4, in which the constellation stars $s_1, \cdots, s_8$ are equally spaced around a circle, with minimum separation $d_{\min}$.

The correct reception probability, assuming that the first symbol is sent, $p(\boldsymbol{x} \in Z_1|m_1$ sent), corresponds to the probability $p(\boldsymbol{x} = \boldsymbol{s}_1 + \boldsymbol{n}|\boldsymbol{s}_1)$. This implies that when noise is added to the transmitted signal $\boldsymbol{s}_1$, the resulting vector $\boldsymbol{x} = \boldsymbol{s}_1 + \boldsymbol{n}$ remains in the shaded region Z_1 shown in the figure.

Figure 6.4 also indicates that the error probability is invariant to any rotation or translation of the signal constellation. The right-hand side of the figure indicates a phase rotation θ, and a translation P of the left-hand side picture. Then, $s_i' = s_i e^{j\theta} + P$.

Expression (6.26) provides an exact solution for the error probability, but it is not possible to determine this probability with a closed function. Therefore, the union bound is generally used to find a solution to that case.

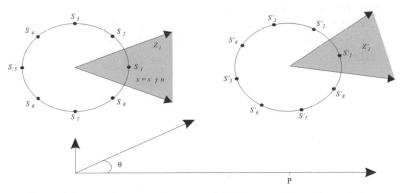

Figure 6.4 Invariance of the error probability to rotation and translation.

Let A_{ik} be the event $||x - s_k|| < ||x - s_i||$, given that the constellation point s_i has been sent. If event A_{ik} occurs, then the signal will be decoded with an error, since the received, once the transmitted constellation s_i is not the nearest point to the received vector x.

On the other hand, A_{ik} does not necessarily implies that s_k is decoded, instead of s_i, because it is possible to exist another constellation point s_j with $||x - s_j|| < ||x - s_k|| < ||x - s_i||$.

The constellation can be correctly decoded if $||x - s_i|| < ||x - s_k|| \ \forall k \neq i$. Then,

$$P_e(m_i \text{ sent}) = p \left(\bigcup_{\substack{k=1 \\ k \neq i}}^{M} A_{ik} \right) \leq \sum_{k=1}^{M} p(A_{ik}), \tag{6.27}$$

in which the inequality follows from the union bound.

Consider $p(A_{ik})$ in detail. This gives,

$$\begin{aligned} p(A_{ik}) &= p(||s_k - x|| < ||s_i - x|| \ |s_i \text{ sent}) \\ &= p(||s_k - (s_i + n)|| < ||s_i - (s_i + n)|| \\ &\quad + p(||n + s_i - s_k|| < ||n||), \end{aligned} \tag{6.28}$$

that is, the error probability equals the likelihood that the noise n is closer to the vector $s_i - s_k$ than to the origin.

The probability depends only on the projection of the noise n on the line that connects the origin and the point $s_i - s_k$, as shown in Figure 6.5. Given the properties of n, that projection is a uni-dimensional Gaussian random variable n, with zero mean and variance $N_0/2$.

Event A_{ik} occurs if n is closer to $s_i - s_k$ than to zero, that is, if $n > d_{ik}/2$, in which $d_{ik} = ||s_i - s_k||$ is the distance between the constellation stars s_i ans s_k. Therefore,

$$p(A_{ik}) = p(n > d_{ik}/2) = \int_{d_{ik}/2}^{\infty} \frac{1}{\sqrt{\pi N_0}} \exp\left[\frac{-v^2}{N_0} \right] dv = Q\left(\frac{d_{ik}}{\sqrt{2N_0}} \right). \tag{6.29}$$

Substitution into Equation (6.27) gives

$$P_e(m_i \text{ sent}) \leq \sum_{\substack{k=1 \\ k \neq i}}^{M} Q\left(\frac{d_{ik}}{\sqrt{2N_0}} \right). \tag{6.30}$$

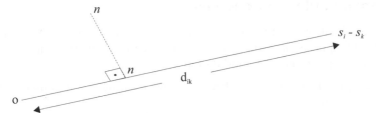

Figure 6.5 Noise projection.

Adding all possible messages,

$$P_e = \sum_{i=1}^{M} p(m_i) P_e(m_i \text{ sent}) \leq \frac{1}{M} \sum_{i=1}^{M} \sum_{\substack{k=1 \\ k \neq i}}^{M} Q\left(\frac{d_{ik}}{\sqrt{2N_0}}\right). \qquad (6.31)$$

Defining the minimum constellation distance as $d_{\min} = \min\limits_{i,k} d_{ik}$, it is possible to simplify Expression (6.31) using the union bound

$$P_e \leq (M-1)Q\left(\frac{d_{\min}}{\sqrt{2N_0}}\right). \qquad (6.32)$$

The use of a well-known bound $Q(x) \leq \frac{1}{2}e^{-x^2/2}$ gives a closed formula for the error probability

$$P_e \leq \frac{M-1}{\sqrt{\pi}} \exp\left[\frac{-d_{\min}^2}{N_0}\right]. \qquad (6.33)$$

For the binary case, $M = 2$, there is only one way to commit an error, and $d_{\min}$ is the distance between two points in the signal constellation; then the upper bound on the Expression (6.32) is exact,

$$P_b = Q\left(\frac{d_{\min}}{\sqrt{2N_0}}\right). \qquad (6.34)$$

Note that P_e is the error probability of a symbol (message), $P_e = p(\hat{m} \neq m_i | m_i \text{ sent})$, in which m_i corresponds to a message with $\log_2 M$ bits, and it is possible to approximate the bit error probability as

$$P_b \approx \frac{P_e}{\log_2 M}, \qquad (6.35)$$

using the Gray code (Lee, 1986).

6.3 Digital Modulation Schemes

The digital modulation schemes use the information bits to select the appropriate amplitude, phase, or frequency, or a combination of them, to modulate the carrier. There are three man modulation systems:

1. Pulse amplitude modulation (PAM) – The information modulates the carrier amplitude.
2. PSK – The signal bits modulate the carrier phase.
3. QAM – The information signal modulates both the carrier amplitude and phase.

Using the notation $s(t) = \Re\{u(t)e^{j2\pi f_c t}\}$ for the transmitted signal, then the modulating signal is given by

$$u(t) = \sum_n s_n g(t - nT_s), \qquad s_n = a_n + jb_n, \qquad (6.36)$$

in which $T_s \gg 1/f_c$ is the symbol interval, n is the interval signaling index, $g(t)$ is the pulse format, and s_n is a complex number that represents $K = \log_2 M$ information bits, that is invariable during the symbol interval.

The bit rate corresponds to K bits per symbol, or $R = K/T_s$ bits second. Because $s_n = a_n + jb_n$ is complex, the signal envelope is variable. The information mapping on the complex number and the choice of the pulse format specify the digital modulation technique.

Therefore, the transmitted signal is

$$s(t) = \left[\sum_n a_n g(t - nT_s)\right] \cos(2\pi f_c t)$$

$$- \left[\sum_n b_n g(t - nT_s)\right] \sin(2\pi f_c t), \qquad (6.37)$$

in which the first term is the in-phase component , $s_I(t)$, and the second term represents the quadrature component, $s_Q(t)$.

The spectral properties of $s(t)$ and $u(t)$ are defined by the spectral characteristics of $g(t)$. The modulated signal bandwidth is, at least, twice the pulse bandwidth, and the design criteria, to minimize the main lobe bandwidth and the amplitude of the other lobes, to reduce co-channel interference, depend on the pulse format.

The mentioned modulation schemes can be represented by the general scheme of Figure 6.6.

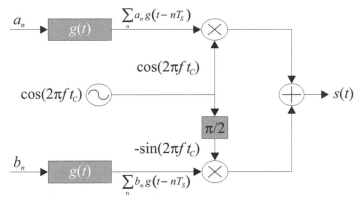

Figure 6.6 General structure for a digital modulator.

6.3.1 Pulse Amplitude Modulation

The simplest modulation system is the M-PAM that has no quadrature component ($b_n = 0$). The information is carried by the amplitude variation, A_m, of the transmitted signal,

$$s_m(t) = \Re \left\{ A_m g(t) e^{j2\pi f_c t} \right\} = A_m g(t) \cos(2\pi f_c t), \qquad 1 \leq t \leq T_s, \tag{6.38}$$

in which $A_m = (2m - 1 - M)d$, $m = 1, 2, \ldots, M$.

The transmitted signal amplitude is obtained from a set of $M = 2^K$ distinct values, which correspond to $\log_2 M = K$ bits per signaling interval T_s.

The transmitted signal energy, in a signaling interval $[0, T_s)$, is

$$
\begin{aligned}
E_{s_m} &= \int_0^{T_s} s_m^2(t)dt = \int_0^{T_s} A_m^2 g^2(t) \cdot \frac{1}{2}[1 + \cos(4\pi f_c t)]dt \\
&\approx \frac{1}{2}A_m^2 \int_0^{T_s} g^2(t)dt = \frac{1}{2}A_m^2 E_g,
\end{aligned} \tag{6.39}
$$

in which E_g is the pulse energy.

This is a good approximation, when $f_c T_s \gg 1$, because $g(t)$ is fairly constant in a cycle $T_c = 1/f_c$. The Euclidean distance between two symbols, which represent different information sequences, is given by

$$d_{mn} = ||s_m(t) - s_n(t)|| = \sqrt{\int_0^{T_s} |s_m(t) - s_n(t)|^2 dt}$$

$$\approx \sqrt{0,5E_g}|A_m - A_n| \geq d\sqrt{2E_g} = d_{\min}, \qquad (6.40)$$

in which $d_{\min}$ is the smallest distance between any two constellation symbols.

It is usual to resort to the Gray code to associate the bit sequences to certain constellation symbols. For this mapping, the adjacent symbols, associated to certain bit sequences, differ in only one bit, as shown in Figure 6.7. Therefore, an error that produces a change from a symbol to an adjacent one only changes a single bit in the sequence.

6.3.2 Phase Shift Keying

For the M-PSK scheme, the information is coded in the transmitted signal phase, that is,

$$
\begin{aligned}
s_m(t) &= \Re\{g(t)e^{j2\pi(m-1)/M}e^{j2\pi f_c t}\}, \qquad 0 \leq t \leq T_s, \\
&= g(t)\cos\left[2\pi f_c t + \frac{2\pi}{M}(m-1)\right] \\
&= g(t)\cos\left[\frac{2\pi}{M}(m-1)\right]\cos(2\pi f_c t) \\
&\quad - g(t)\sin\left[\frac{2\pi}{M}(m-1)\right]\sin(2\pi f_c t).
\end{aligned}
\qquad (6.41)
$$

Thus, for $s_n = a_n + jb_n$, the coefficients a_n and b_n are given as $\cos\left[\frac{2\pi}{M}(m-1)\right]$ and $\sin\left[\frac{2\pi}{M}(m-1)\right]$ respectively. As in the M-PAM case, $g(t)$ is the pulse format, and $\theta_m = \frac{2\pi}{M}(m-1)$, $m = 1, 2, \ldots, M$ are the possible carrier phases that contain the information bits.

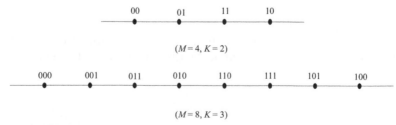

Figure 6.7   Gray mapping for the M-PAM modulation scheme.

The signals have equal energy,

$$E_{s_m} = \int_0^{T_s} s_m^2(t)dt \approx \frac{1}{2}\int_0^{T_s} g^2(t)dt = \frac{1}{2}E_g. \qquad (6.42)$$

Observe that for a rectangular pulse, $g(t) = 1$ in the pulse interval, the transmitted signal has constant envelope, differing from the M-PAM and M-QAM schemes. However, rectangular pulses are spectrally inefficient.

The distance between two M-PSK constellation symbols is given by

$$d_{mn} = \sqrt{E_g\left(1 - \cos\left(\frac{2\pi}{M}(m-1)\right)\right)} \geq \sqrt{E_g(1 - \cos(2\pi/M))} = d_{\min}.$$

$$(6.43)$$

As in the case of M-PAM, it is common to use the Gray code, as illustrated in Figure 6.8.

6.3.3 Quadrature Modulation

For the M-QAM scheme, the information bits modulate both the carrier phase and amplitude. There are, then, two degrees of freedom to code the information bits. As a result, the M-QAM scheme has better spectral efficiency, that is, for a given bandwidth and a certain average power, the M-QAM manages to code a larger number of bits per symbol.

The transmitter signal is

$$s_m(t) = \Re\{A_m e^{j\theta_m}g(t)e^{j2\pi f_c t}\} = A_m g(t)\cos(2\pi f_c t + \theta_m), \quad 0 \leq t \leq T_s.$$

$$(6.44)$$

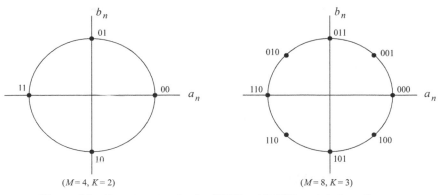

Figure 6.8   Gray mapping for the QPSK and 8-PSK modulation schemes.

The energy of the signal $s_m(t)$ is

$$E_{sm} = \int_0^{T_s} s_m^2(t)dt \approx \frac{1}{2}A_m^2 E_g, \tag{6.45}$$

which is the same result for the M-PAM scheme.

The Euclidean distance between two constellation symbols is given by

$$d_{mn} = \sqrt{\frac{1}{2}E_g[(a_m - a_n)^2 + (b_m - b_n)^2]}, \tag{6.46}$$

in which $a_k = A_k \cos\theta_k$ and $b_k = A_k \operatorname{sen}\theta_k$ for $k = m, n$.

Considering square M-QAM in which a_n and b_n assume values $(2m - 1 - L)d$ for $m = 1, 2, \ldots, L = 2^l$, the smaller distance between the symbols is $d_{\min} = d\sqrt{2E_g}$.

This is equivalent to the distance obtained for the M-PAM constellation. On the other hand, the number of symbols in a square QAM constellation is $M = d^{2l}$. The transmission rate is $2l$ bits/symbol, that is, l bits per dimension. The common square constellations are 4-QAM and 16-QAM, illustrated in Figure 6.9.

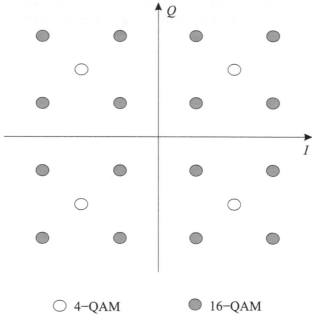

$\bigcirc$ 4–QAM $\bullet$ 16–QAM

Figure 6.9   Constellations 4-QAM and 16-QAM.

6.4 Differential Coding

In the M-PSK and M-QAM schemes, the information modulates the carrier phase; therefore, they require coherent demodulation to recover the original signal. This implies that the carrier phase must be known at the receiver.

The carrier phase can be recovered using a pilot signal, a low-amplitude sub-carrier that is transmitted along with the signal, or using a PLL (Koufalas, 1996). However, the additional complexity raises receiver costs and makes the detection more susceptible to carrier phase drift, for example.

Besides, it is difficult to obtain a phase reference when dealing with fast fading channels. Therefore, differential modulation schemes, that do not require phase references, are usually preferred in wireless applications.

The differential modulation techniques belong to a general modulation class known as modulation with memory, in which a symbol transmitted in a discrete instant n depends either on current and on past bits ($n - 1$, for instance).

Differential modulation can be used to adjust the transmitted signal spectrum format according to the channel spectral characteristics, as in the case of non-return to zero (NRZ) and non-return to zero inverted (NRZI).

But, the main advantage of differential modulation is to avoid the need for a phase reference to demodulate a signal. The principle is to use the previously sent symbol as the phase reference for the current symbol. That is, the information bits code the phase difference between the current and the past symbols.

The differential coding is less sensitive to carrier random phase shifts. However, if the the channel is subject to Doppler shift, for instance, when the receiver is moving, the phases in distinct time intervals can be uncorrelated, which renders the previous symbol a noisy phase reference. This implies that the error probability will have a floor, an irreducible level.

6.5 Offset Phase Modulation

Every baseband signal, $s_n = a_n + jb_n$, that modulates a carrier in phase assumes a phase in certain phase in the four quadrants of the complex plane. At time nT_s, the transition to a new symbol can imply a $180°$ phase change, and this can make the signal amplitude to pass through zero.

Sudden phase changes, and large amplitude variations produce sidelobes, and can be distorted by amplifiers and filters. The abrupt transitions can be avoided with the introduction of an offset in the quadrature pulse, which is accomplished by the introduction of a delay of a half symbol period, as shown in Figure 6.10.

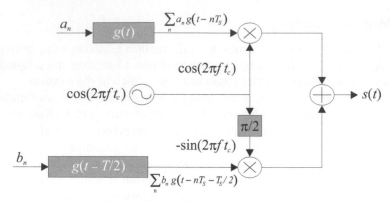

Figure 6.10   Offset phase modulator.

The offset quadrature phase modulation is usually denoted as OQPSK, a scheme that has the same spectral characteristics of the QPSK system, for linear amplification, but has a better efficiency for non-linear operation, because the maximum phase transition is $90°$.

Another technique to minimize the phase transitions, used in some cellular mobile standards, is the $\pi/4$-QPSK, that rotates the QPSK by $45°$, and avoids switching the oscillators on and off. It can also be deferentially coded, eliminating the need for a phase reference, producing the $\pi/4$-DQPSK.

6.6 The Transmission Pulse

For the M-PSK scheme, if $g(t)$ is a rectangular pulse of width T, the signal envelope is constant, which gives an advantage regarding non-linear amplification, such as obtained from a satellite transponder amplifier. But, rectangular pulses, as demonstrated in Chapter 2, have high-energy sidelobes, which obliges the designer to use a large bandwidth, in order to minimize co-channel interference.

Of course, the pulse format can be modified, to reduce the out-of-band energy, but this must be done without introducing inter-symbol interference (ISI). Therefore, the pulse format $g(t)$ must satisfy the Nyquist criterion, which states that the the ISI must be null at the sampling instants (Haykin, 1989).

The following pulse formats satisfy the Nyquist criterion:

1. Rectangular pulse – That is, $g(t) = 1, 0 \leq t \leq T_s$. This pulse format produces a constant envelope, for M-PSK schemes, but has poor spectral properties, with high-energy sidelobes.

2. Sine or cosine pulse – That is, $g(t) = \sin(\pi t/T_s)$, $0 \le t \le T_s$. This pulse format is often used in the minimum shift keying (MSK) system. The quadrature symbol is delayed by $T_s/2$, and this produces a constant amplitude modulation and low-energy sidelobes. The Gaussian minimum shift keying (GMSK) modulation scheme, used in the Global System for Mobile Communications (GSM) standard, has this pulse format.

3. Raised cosine pulse – It has been designed in the frequency domain, to obtain the desired spectral properties. Thus, the pulse $g(t)$ is specified as a function of its Fourier transform,

$$G(f) = \begin{cases} T_s, \ 0 \le |f| \le (1-\beta)/2T_s, \\ \frac{T_s}{2}\left[1 - \sin\left(\frac{\pi T_s}{\beta} \cdot \left(f - \frac{1}{2T_s}\right)\right)\right], \\ (1-\beta)/2T_s < |f| \le (1+\beta)/2T_s, \end{cases} \quad (6.47)$$

in which β is defined as the roll-off factor that determines the spreading percentage of the pulse.

The time domain representation of pulse $g(t)$, corresponding to $G(f)$, is given by

$$g(t) = \frac{\text{sen}(\pi t/T_s)}{\pi t/T_s} \frac{\cos(\beta \pi t/T_s)}{1 - 4\beta^2 t^2/T_s^2}. \quad (6.48)$$

Figures 6.11 and 6.12 present the raised cosine pulse in the time and frequency domains, for some values of β. For $\beta = 0$, the pulse is rectangular. The pulse amplitude decay, in the time domain, is proportional to $1/t^3$, faster than the previous ones.

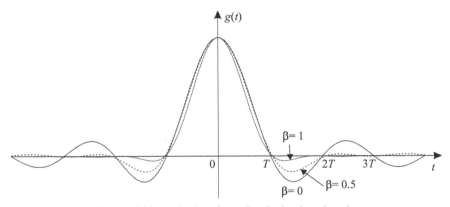

Figure 6.11 Raised cosine pulse, in the time domain.

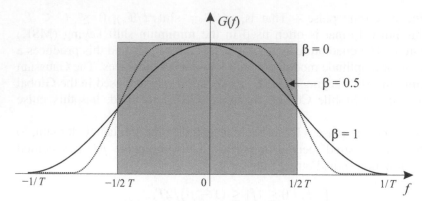

Figure 6.12   Raised cosine pulse, in the frequency domain.

6.7 Constant Envelope Modulation

Constant envelope modulation permits the use of non-linear amplification, that is usually more efficient in terms of energy use. It is also more robust to channel fading and noise. However, this type of modulation has poor spectral efficiency, and occupies a larger bandwidth. Frequency shift keying (FSK) is a constant envelope modulation scheme, which has the following variations: M-FSK, MSK, and GMSK.

For the M-FSK scheme, the modulated signal has the format given by

$$s_m(t) = A\cos(2\pi f_c t + 2\pi \alpha_m \Delta f_c t), \qquad 0 \le t \le T_s, \qquad (6.49)$$

in which $\alpha_m = (2m - 1 - M)$, $m = 1, 2, \ldots, M = 2^K$. This way, the smaller frequency distance between different symbols is $2\Delta f_c$.

The usual way to generate the M-FSK signal is to change the carrier frequency directly, as in the analog FM case,

$$s_m(t) = A\cos\left(2\pi f_c t + 2\pi \beta \int_{-\infty}^{t} u(\tau)d\tau\right) = A\cos(2\pi f_c t + \theta(t)),$$
$$(6.50)$$

in which $u(t) = \sum_n a_n g(t - nT_S)$ is an M-PAM signal modulated by the information bits, as described previously. The phase $\theta(t)$ is continuous, and the scheme is called continuous phase frequency shift keying (CPFSK).

Using Carson's rule, for small values of β, gives the following bandwidth for $s(t)$

$$B_s \approx M\Delta f_c + 2B, \qquad (6.51)$$

in which B is the M-PAM signal bandwidth ($u(t)$).

As discussed, the MSK modulation is a special case of FSK, in which the frequency separation is $\Delta f_c = 0, 5/T_s$. This is the minimum frequency separation, for which the inner product $< s_m(t), s_n(t) >$ is null in a symbol interval, for $m \neq n$. Therefore, the MSK scheme occupies the smaller bandwidth.

The pulse format that is more commonly used to improve the MSK spectral efficiency is the Gaussian one,

$$g(t) = \frac{\sqrt{\pi}}{\alpha} e^{-\pi^2 t^2/\alpha^2},\qquad(6.52)$$

in which α is a parameter related to the spectral efficiency of the system.

The Gaussian pulse spectrum is given by

$$G(f) = e^{-\alpha^2 f^2}\qquad(6.53)$$

in which the parameter α is related to the bandwidth,

$$\alpha = \frac{\sqrt{-\ln\sqrt{1/2}}}{B}.\qquad(6.54)$$

An increase in the value of α improves the spectral efficiency of the GMSK scheme.

6.8 The Rotated Constellation

This section presents a performance analysis of a modified QPSK scheme for transmission over fading channels. Interleaving and rotation of the signal constellation are applied with the objective of improving its performance. It is shown that the proposed system outperforms the original scheme even in the presence of channel estimation errors.

Fading causes significant degradation in the performance of digital wireless communications systems. Unlike the additive Gaussian channel, the wireless channel suffers from attenuation due to destructive addition of multi-paths in the propagation media and due to interference from other users (Tarokh et al., 1998). In order to minimize the effects of fading, some systems use a resource called diversity which consists, basically, in providing replicas of the transmitted signals at the receiver. Examples of diversity techniques are: temporal diversity, frequency diversity, and antenna diversity.

Another way to increase the system diversity is to introduce redundancy by combining rotation and interleaving of the constellation symbols before the modulation process (Divsalar and Simon, 1988; Kerpez, 1993; Jeličić and

Roy, 1995; Boutros and Viterbo, 1998; Slimane, 1998). This recent technique has been called modulation diversity.

A considerable performance gain, in terms of the bit error probability, can be achieved by selecting the rotation angle, for a QPSK constellation, when the channel is modeled by Rayleigh fading. However, it has been assumed ideal channel state information (ideal CSI) at the receiver, which is a very restrictive assumption for practical systems where channel estimation errors should be considered. This section presents the performance of the system proposed in (Lopes and Alencar, 2000) in the absence of ideal CSI, that is, considering the presence of channel estimation errors.

6.8.1 The Modulation Diversity Technique

The key point to increase the modulation diversity is to apply a certain amount of rotation to a classical signal constellation in such a way that any two points achieve the maximum number of distinct components (Boutros and Viterbo, 1998).

Figure 6.13 illustrates this idea for a QPSK scheme. In fact, if it is supposed that a deep fade hits only one of the components of the transmitted signal vector, then one can see that the compressed constellation in Figure 6.14 (empty circles) offers more protection against the effects of noise, since no two points collapse together as it would happen in Figure 6.13. A component interleaver/deinterleaver is required to assume that the in-phase and quadrature components of the received symbol are affected by independent fading.

An interesting feature of the rotation operation is that the rotated signal set has exactly the same performance of the non-rotated one when used over a

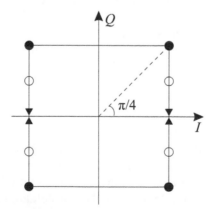

Figure 6.13   The original constellation.

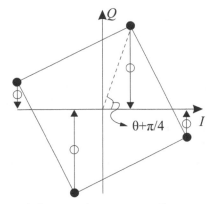

Figure 6.14   The performance improvement using a rotated constellation.

pure AWGN. This occurs due to the fact that the Euclidean distance between the symbols, for rotated and non-rotated QPSK constellations, is the same in the two cases.

6.8.2 Rotating the QPSK Constellation

For a frequency-nonselective slowly fading channel, coded modulation combined with interleaving showed good performance improvement of digital communications systems. Interleaving destroys fading correlation, adding diversity to the coded scheme (Divsalar and Simon, 1988; Jeličić and Roy, 1995).

In (Jeličić and Roy, 1995), the author used a new form of interleaving, in which each coordinate is interleaved independently to enhance the system performance. In (Lopes and Alencar, 2000), the key idea was to analyze the influence of rotating the QPSK constellation, by a constant phase θ, in the system performance. The block diagram of the modified QPSK scheme transmitter is shown in Figure 6.15, and the receiver is shown in Figure 6.16.

Consider a conventional QPSK scheme. The transmitted signal is given by

$$
\begin{aligned}
s(t) = A \sum_{n=-\infty}^{+\infty} a_n p(t - nT_S) \cos(2\pi f_c t) \\
+ A \sum_{n=-\infty}^{+\infty} b_n p(t - nT_S) \sin(2\pi f_c t)
\end{aligned}
\tag{6.55}
$$

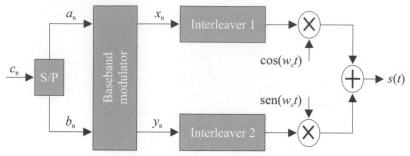

Figure 6.15   Block diagram of the modified QPSK transmitter.

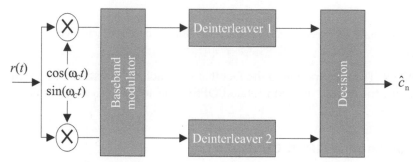

Figure 6.16   Block diagram of the modified QPSK receiver.

where

$$a_n, b_n = \pm 1 \quad \text{with equal probability,}$$

$$p(t) = \begin{cases} 1, & 0 \le t \le T_S \\ 0, & \text{elsewhere,} \end{cases}$$

f_c is the carrier frequency, and A is the carrier amplitude.

When the signals in phase (I channel) and quadrature (Q channel) are interleaved independently, a diversity gain can be obtained because the fading in one channel is independent from the other channel. From Equation (6.55), the sequence $\{a_n\}$ is independent of the sequence $\{b_n\}$. In this case, the system cannot take advantage of the previous described diversity unless some kind of redundancy between the two quadrature channels is introduced. Introducing redundancy in the QPSK scheme can be achieved by rotating its signal constellation by a constant phase θ, as shown in Figure 6.17.

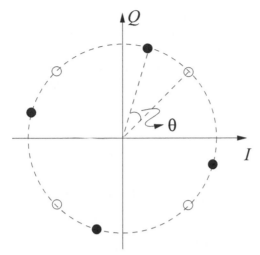

Figure 6.17 The rotated QPSK constellation.

After the rotation and the interleaving, the transmitted signal becomes

$$s(t) = A \sum_{n=-\infty}^{+\infty} x_n p(t - nT_S) \cos(2\pi f_c t)$$

$$+ A \sum_{n=-\infty}^{+\infty} y_{n-k} p(t - nT_S) \sin(2\pi f_c t),$$

(6.56)

in which k is an integer representing the time delay, in number of symbols, introduced by the interleaving between the I and Q components and

$$x_n = a_n \cos\theta - b_n \sin\theta$$
$$y_n = a_n \sin\theta + b_n \cos\theta$$

(6.57)

are the rotated symbol components. It is important to note that the rotation does not affect the system spectral efficiency for the rotated system that also transmits two bits in one symbol interval.

The digital communication channel is assumed to be frequency-non-selective slowly fading with a multiplicative factor representing the effect of fading and an additive term representing the AWGN channel. It is also assumed that the system is unaffected by intersymbol interference. Thus, the received signal is then written as

$$r(t) = \alpha(t)s(t) + \eta(t)$$

(6.58)

in which $\alpha(t)$ is modeled as zero-mean complex Gaussian process.

If coherent demodulation is used, the fading coefficients can be modeled, after phase elimination, as real random variables with Rayleigh distribution and unit second moment $(E[\alpha^2] = 1)$. The independence of the fading samples represents the situation where the components of the transmitted points are perfectly interleaved. An undesirable effect of the component interleaving is the fact that it produces a non-constant envelope transmitted signals (Boutros and Viterbo, 1998).

6.8.3 The Presence of Channel Estimation Errors

If channel estimation errors are considered, the receiver is not able to obtain the correct value of the fading coefficient α in the decoding process; thus, the estimation $\beta(t)$ of the fading coefficient used by the decoder can be written as

$$\beta(t) = \alpha(t) + \delta(t), \tag{6.59}$$

in which $\delta(t)$ represents the channel estimation error.

In this work, $\delta(t)$ is modeled as a complex Gaussian random variable having zero-mean. Regardless of which method is used for estimation, the variance of $\delta(t)$ can be directly obtained from the Cramér–Rao Bound (Tarokh et al., 1999), which establishes that the variance of the estimation error is

$$\text{Var}[\delta(t)] = \frac{No}{2KE_s}, \tag{6.60}$$

in which K is the number of pilot symbols and E_s is the symbol energy. As expected, it can be seen from the previous equation that the variance of the estimation error decreases when the SNR increases.

After channel estimation, the decoder selects, for each received signal, the symbol constellation $s(t)$ that minimizes the metric

$$d^2 = |r(t) - \beta(t)s(t)|. \tag{6.61}$$

6.8.4 Simulation Results

This section presents the simulations used to find out the rotation phase which produces the best performance, in terms of BER, considering no channel estimation errors (ideal CSI). Next, this rotation is applied to the scheme considering the estimation errors and its performance is assessed by simulations.

In order to determine the phase rotation θ that achieves the best performance, the transmission system in Figure 6.15 was simulated for θ varying

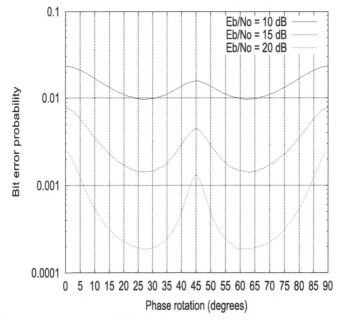

Figure 6.18 The bit error probability for the modified QPSK scheme as a function of the phase rotation θ. Ideal channel state information (CSI) was considered.

from zero to $\pi/2$. Figure 6.18 shows the system performance, measured in terms of bit error probability, for $E_b/N_0 = 10$, 15, and 20 dB.

It can be seen that the optimum performance is obtained for θ approximately equal to $\pi/7$ for the three curves presented. Considering this optimum phase rotation, Figure 6.19 compares the performance of the original QPSK scheme ($\theta = 0$) and its rotated version for E_b/N_0 varying from zero to 30 dB. When $\theta = 0$, the performance of the system reduces to that of a conventional QPSK scheme. It can be noted that a considerable performance improvement is obtained compared to the conventional QPSK scheme, which can reach 8 dB at a bit error probability of 10^{-3}.

Figure 6.19 also shows the performance of the proposed system in the absence of ideal channel estimation. The dotted lines represent the bit error probability for the original system and the rotated system. It is important to note that, in both cases, the estimation errors increase the bit error probability.

However, the rotated system is more robust than the original scheme in terms of performance decrease. For example, at a bit error probability of 10^{-2}, the channel errors produce a performance decrease of 1 dB, for the rotated system, while this drop can reach 5 dB for the non-rotated system.

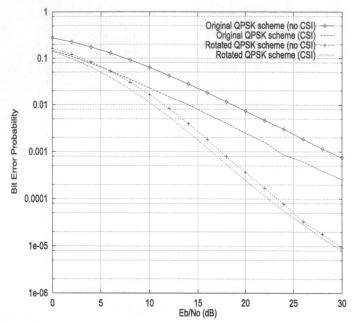

Figure 6.19 The bit error probability for the rotated QPSK scheme and the non-rotated QPSK scheme as a function of E_b/N_0. Two cases were considered: Ideal CSI and absence of CSI.

This section presents the performance analysis of a rotated-based system in the absence of ideal CSI. The effects of the presence of channel estimation errors were considered. As can be inferred from the simulation results, the optimum performance is obtained for a rotation angle θ approximately equal to $\pi/7$ for the three curves presented. At this rotation phase, a considerable performance improvement, over the original QPSK scheme, is obtained.

6.9 Orthogonal Frequency Division Multiplexing

The orthogonal frequency division multiplexing (OFDM) scheme has been chosen as the transmission scheme for most digital television, digital radio, and wireless standards.

This section presents an stochastic analysis of the OFDM signal, to obtain general expressions for its autocorrelation and PSD formulas. The resulting PSD models exactly what is produced by practical spectrum analyzers.

OFDM is a variant of a technique known as frequency division multiplexing (FDM), which was widely used, for many years, by telecommunication companies in analog systems for long distance transmission. It is also known as discrete multi-tone modulation (DMT) and aims at transmitting multiple signals in different frequencies (Alencar, 2001).

The OFDM scheme uses channel coding, which is a technique for correcting transmission errors, giving rise to the acronym COFDM. The technique typically requires a fast fourier transform (FFT), which is implemented in a digital signal processor (DSP) or directly in an integrated circuit (IC) (Rajanna et al., 2011).

The OFDM has been used in radio broadcasting, communication networks, and computer networks. Its main performance attribute is its robustness to multipath effect, which leads to the ghost effect in the reception with analog television, as well as to the frequency selective fading, in mobile communications (Alencar and da Rocha Jr., 2005).

An OFDM baseband signal is a result from the combination of many orthogonal subcarriers, with the data of each subcarrier independently modulated with the use of any QAM or PSK technique. This signal is used to modulate a main carrier for radiofrequency broadcasting.

Some advantages come from the use of OFDM, including high spectral efficiency, robustness against multipath and burst noise, as well as robustness to slow phase distortion and fading. This is achieved by combining OFDM with techniques of error correction, adaptive equalization, data interleaving, and reconfigurable modulation. Additionally, the COFDM scheme has a spectrum which is typically uniform, which leads to some advantages regarding cochannel interference (van Waterschoot et al., 2010).

6.9.1 Description of OFDM

The ISDB-T uses the OFDM technique as a transmission scheme. The subcarriers form a set of functions that are orthogonal to each other, that is, the integral of the product between any two of these functions within the interval of a symbol is null.

This orthogonality ensures that the ISI in the frequencies of the subcarriers be null. Figure 6.20 illustrates the effect of the orthogonality. On the other hand, the orthogonality also allows for the narrowest possible band occupied by the OFDM signal, which, in turn, makes the signal fit into a 6-MHz passband channel.

In Figure 6.20, the frequency of the main subcarrier (f_c) equals the inverse of the duration of the symbol ($\frac{1}{T_S}$). In the 8K mode, f_c is 837,054 Hz and, in the 2K mode, it is equal to 3,348,214 Hz. The values calculated for f_c result from the need to maintain the orthogonality between the subcarriers.

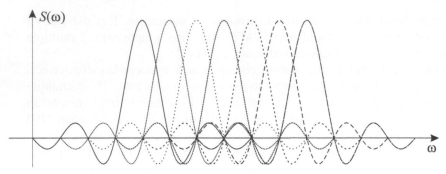

Figure 6.20   Orthogonality between the carriers of the OFDM system.

Each subcarrier is modulated in QPSK, 16-QAM, or 64-QAM by one of the v bit sets mapped by the block mapper. For every set of v bits (2, 4, or 6 bits) there is a given state of phase/amplitude of the subcarrier. During a symbol (T_S), the states of the subcarriers remain unchanged. In the next symbol, they will acquire new states due to the new sets of v bits that are found in the input of the modulators of each subcarrier.

It is worth mentioning that the state of the subcarriers, within the transmission of a symbol, possesses the information of the frequency spectrum isolated from the OFDM signal. To convert this information to time domain, the IFFT is used. All these operations are digitally performed by means of digital processors. The obtained OFDM signal is in digital format and ready to be injected in the next block, in which the guard interval is inserted (Skrzypczak et al., 2006).

Figure 6.21 shows the difference between the transmission techniques with single carrier, used in the Advanced Television Systems Committee (ATSC) standard, and multicarrier, used in the Digital Video Broadcasting (DVB), Integrated Services Digital Broadcasting (IDSB), and ISDB-Tb (*Brazilian Digital Television System*) standards.

The figure shows the spectra $S(\omega)$, in which ω_k is the frequency of the k-th subcarrier (in radians per second), W_{SC} is the passband for the signal modulated with single carrier, and W_{MC} is the total passband for a transmission with N carriers (Alencar, 2012).

Defining the spectrum of each deterministic modulated carrier in OFDM by $F_k(\omega)$, the spectrum of the composed signal is given by

$$S(\omega) = \sum_{k=1}^{N} F_k(\omega). \tag{6.62}$$

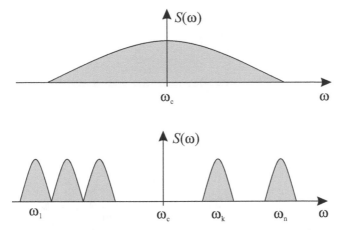

Figure 6.21 Transmission techniques: (a) Single carrier and (b) OFDM.

As a consequence, the transmitted signal is written as

$$s(t) = \sum_{l=-\infty}^{\infty} \sum_{k=1}^{N} b_{kl} e^{j\omega_k(t-lT_S)} f(t - lT_S), \qquad (6.63)$$

in which b_{kl} is the l-th information *bit* in the k-th subcarrier, $f(t)$ represents the pulse waveform for the transmitted signal, and T_S is the symbol interval.

By taking discrete samples of the signal $s(t)$, the resulting equation represents the inverse discrete Fourier transform of the signal. Hence, for demodulating the OFDM signal, it suffices to obtain the direct Fourier transform of the received signal.

This is useful for a preliminary analysis, but does not represent the measurement of a real signal. First, there is no negative power, as most engineers know. Second, real signals are random, not deterministic. Therefore, it is important to deal with practical signals and PSDs, instead of Fourier transform of deterministic waveforms.

The power spectrum density is useful to analyze the channel transmission effects. For example, a multipath channel, which is typical in digital television transmission, leads to a selective attenuation in some range of frequencies, affecting the signal reception. Figure 6.22 shows the effect of the channel filtering, caused by multipath, for instance, in the transmission with single carrier and OFDM.

As can be observed in Figure 6.22, the effect of the attenuation caused by the channel transfer function $H(\omega)$ is smaller for the OFDM transmission, since it affects only some subcarriers, whose information rates are slow. As a

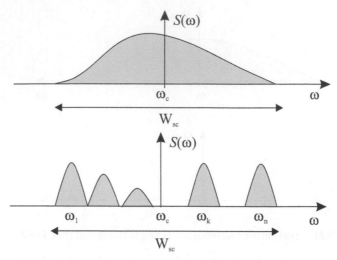

Figure 6.22   Channel effect: (a) Single carrier and (b) OFDM.

consequence, the received signal does not need equalization or it needs only simple equalization for each subcarrier. The single carrier system, by its turn, requires a complex adaptive equalization subsystem.

The OFDM signal is affected by fast time-variant channels or by the frequency offset of some carrier. Additionally, the signal tends to have high peak-to-average ratio.

The OFDM technique is used in ADSL (Asymmetric Digital Subscriber Line) connections that follows the G.DMT (ITU G.922.1) standard and is also employed in wireless local area networks, including the IEEE 802.11a/g standard, or HIPERLAN/2, and the data wireless transmission WiMax system. This technique is used in Europe and in other localities in which the standards Eureka 147 Digital Audio Broadcast (DAB) and Digital Radio Mondiale (DRM) were adopted for digital radio transmission as well as for television transmission in DVB standard.

For digital television transmission, one argues that the COFDM systems, used in Japanese and European standards, have robustness to multipath higher than the 8-VSB, used in the American standard. The first 8-VSB digital television receivers had signal reception problems in urban areas; however, most recent receivers have evolved in this aspect. The 8-VSP modulation requires a smaller transmitted power, which is interesting for small cities. However, in urban areas or with geographical accidents, the COFDM provides a better reception (Alencar, 2009a).

Generally, the TV systems uses 2,000 to 8,000 carriers, which can be modulated with QPSK (quadrature phase shift keying), 16-QAM, or

64-QAM. As a consequence, each carrier transports a relatively low bit rate. Besides, as each part of the signal is transported by a carrier in a different frequency, this provides certain noisy immunity with respect to interference in specific frequencies, since only part of the information is affected.

The modulated symbols in QAM or QPSK are transformed with the inverse fast Fourier transform (IFFT) and positioned in the time domain in orthogonal carriers. Upon receiving the signal, the receiver must only perform the FFT in the clocks of received signal blocks to obtain the transmitted signal. Guard intervals are inserted between the bits.

6.9.2 COFDM Transmission

The terrestrial channel is different from satellite and cable channels, and is exposed to multipath effects, as a result of the surface topography and the presence of buildings. During the DVB-T development, the mobile reception was not an exigency, but the system can handle the signal variation due to the movement of transmitters/receivers.

The system that presents the best performance for such channel conditions is the COFDM. COFDM manages the multi-path problem very well, and offers a high level of frequency economy, allowing a better utilization of the available bandwidth (Ladebusch and Liss, 2006). The COFDM technique is based on the utilization of several carriers which transport the signal in FDM sub-channels with 6, 7, or 8 MHz.

In this way, each carrier transports a limited bit rate. The interference between the carriers is avoided by making them orthogonal. That occurs when the separation between the carriers is the inverse of the period over which the receiver will accomplish the signal demodulation (Reimers, 1997).

In the DVB standard, 1,705 (in the 2K mode) or 6,817 (in the 8K mode) carriers are used. To increase the immunity to external interferences, several codification techniques are employed, which includes a pseudo-random exchange of the useful load among the carriers.

For a given configuration of the above-mentioned parameters, the bits are assembled to form a word. Each word modulates a carrier, during the time of one symbol (T_S). The set of words of all the carriers in a given time interval T_S is called a COFDM symbol.

Thus, the COFDM symbol defined for the DVB-T consists in any of the 6,817 carriers of the 8K mode or the 1,705 carriers of the 2K mode. In the DVB-T system, the OFDM symbols are combined to form a transmission frame, which consists in 68 consecutive symbols. Four consecutive transmission frames form a super-frame (Ladebusch and Liss, 2006).

The beginning of each COFDM symbol is preceded by an interval called guard band. The aim of this interval is to guarantee the immunity to echoes and reflections, so that if, for example, an echo is in the guard band, it does not affect the decoding of useful data transmitted on the symbols. The guard band is a cyclic continuation of the symbol and its length in relation to the duration of a symbol can assume four different values: 1/4, 1/8, 1/16, or 1/32 of symbol time (Panta and Armstrong, 2003).

To achieve synchronism and phase control, pilot-signals are inserted. These pilots are transmitted with an amplitude 1/0.75 larger than the amplitude of the others carriers, with the aim of making them particularly robust to the transmission effects. The two end carriers (0 and 1,704 for the 2K mode and 0 and 6,816 for the 8K mode) play this role. Another 43 carriers in the 2K mode and 175 for the 8K mode are used as continuous pilot.

For the remaining carriers, some words are used in predefined sequences to act as dispersed pilot signals, which are used to estimate the transmission carrier (and neighboring carriers) characteristics (Yamada et al., 2004). Some carriers are used to send a TPS control signal which identifies the transmission channel parameters, such as modulation type and number of carriers.

6.9.3 OFDM with Random Signals

A more general expression for the PSD of the orthogonal frequency division scheme can be obtained using the theory of modulation with stochastic processes. The resulting power spectrum represents the very signal which the engineer observes when operating a spectrum analyzer.

The OFDM signal is the sum of randomly modulated signals

$$s(t) = \sum_{k=1}^{N} s_k(t), \tag{6.64}$$

in which the quadrature modulated signals $s_k(t)$ can be written as

$$s_k(t) = b_k(t)\cos(\omega_k t + \phi) + d_k(t)\sin(\omega_k t + \phi). \tag{6.65}$$

The random modulating signals $b(t)$ and $d(t)$ can be correlated or uncorrelated and the phase ϕ is a random variable uniformly distributed in a interval of 2π radians and independent of $b(t)$ and $d(t)$.

6.9.4 Quadrature Modulation with Random Signals

The QAM is used in conjunction with the OFDM technique It uses sine and cosine orthogonality properties to allow the transmission of two different signals in the same carrier, which occupies a bandwidth that is equivalent to the AM signal. The information is transmitted by both the carrier amplitude and phase.

It is possible to write each modulated signal as

$$s_k(t) = \sqrt{b_k^2(t) + d_k^2(t)} \cos\left(\omega_k t - \tan^{-1}\left[\frac{d_k(t)}{b_k(t)}\right] + \phi\right),\qquad(6.66)$$

in which the modulating signal, or amplitude resultant, can be expressed as $a_k(t) = \sqrt{b_k^2(t) + d_k^2(t)}$ and the phase resultant is $\theta_k(t) = -\tan^{-1}\left[\frac{d_k(t)}{b_k(t)}\right]$.

The autocorrelation function for the composite OFDM signal can be computed using the definition of the expected value operator $E[\cdot]$,

$$R_S(\tau) = E[s(t) \cdot s(t + \tau)].\qquad(6.67)$$

Substituting Equations (6.64) and (6.65) into (6.67) gives

$$R_S(\tau) = E\left[\sum_{k=1}^{N} b_k(t)\cos(\omega_k t + \phi) + d_k(t)\sin(\omega_c t + \phi)\right.$$

$$\cdot\left[\sum_{i=1}^{N} b_i(t+\tau)\cos(\omega_c(t+\tau) + \phi)\right.$$

$$\left.+ \sum_{i=1}^{N} d_i(t+\tau)\sin(\omega_i(t+\tau) + \phi)\right].\qquad(6.68)$$

Expanding the product, and using the linearity property of the expected value, gives

$$R_S(\tau) = \sum_{k=1}^{N}\sum_{i=1}^{N} E[b_k(t)b_i(t+\tau)\cos(\omega_k t + \phi)\cos(\omega_i(t+\tau) + \phi)$$

$$+ d_k(t)d_i(t+\tau)\sin(\omega_k t + \phi)\sin(\omega_i(t+\tau) + \phi)$$

$$+ b_k(t)d_i(t+\tau)\cos(\omega_k t + \phi)\sin(\omega_i(t+\tau) + \phi)$$

$$+ b_i(t+\tau)d_k(t)\cos(\omega_i(t+\tau) + \phi)\sin(\omega_k t + \phi)].\qquad(6.69)$$

Using trigonometric and orthogonality properties, and collecting terms which represent known autocorrelation functions, it follows that

$$R_S(\tau) = \sum_{k-1}^{N}\left[\frac{R_{B_k}(\tau) + R_{D_k}(\tau)}{2}\right]\cos\omega_k\tau$$

$$+ \left[\frac{R_{BD_k}(\tau) - R_{DB_k}(\tau)}{2}\right]\sin\omega_k\tau.\qquad(6.70)$$

Considering zero-mean uncorrelated modulating signals,

$$R_{BD_k}(\tau) = E[b_k(t)d_k(t+\tau)] = 0$$

and

$$R_{DB_k}(\tau) = E[b_k(t+\tau)d_k(t)] = 0,$$

the resulting autocorrelation is then given by

$$R_S(\tau) = \sum_{k=1}^{N} \frac{R_{B_k}(\tau)}{2} \cos \omega_c \tau + \frac{R_{D_k}(\tau)}{2} \cos \omega_c \tau. \qquad (6.71)$$

The carrier power is given by the following formula

$$P_S = R_S(0) = \sum_{k=1}^{N} \frac{P_{B_k} + P_{D_k}}{2}. \qquad (6.72)$$

Because of the linearity of the OFDM technique, computing the autocorrelation of $s(t)$ in Equation (6.64), and Fourier transforming the resulting formula, one obtains the power spectrum density of the OFDM composite signal,

$$
\begin{aligned}
S_S(\omega) &= \frac{1}{4} \left[\sum_{k=1}^{N} S_{B_k}(\omega + \omega_k) + S_{B_k}(\omega - \omega_k) \right] \\
&+ \frac{1}{4} \left[\sum_{k=1}^{N} S_{D_k}(\omega + \omega_k) + S_{D_k}(\omega - \omega_k) \right] \\
&+ \frac{j}{4} \left[\sum_{k=1}^{N} S_{BD_k}(\omega + \omega_k) - S_{BD_k}(\omega - \omega_k) \right] \\
&+ \frac{j}{4} \left[\sum_{k=1}^{N} S_{DB_k}(\omega - \omega_k) - S_{DB_k}(\omega + \omega_k) \right], \qquad (6.73)
\end{aligned}
$$

in which $S_{B_k}(\omega)$ and $S_{D_k}(\omega)$ represent the respective power spectrum densities for $b_k(t)$ and $d_k(t)$; $S_{BD_k}(\omega)$ is the cross-spectrum density between $b_k(t)$ and $d_k(t)$; $S_{DB_k}(\omega)$ is the cross-spectrum density between $d_k(t)$ and $b_k(t)$.

For uncorrelated signals, the previous formula can be simplified to

$$S_S(\omega) = \frac{1}{4}\left[\sum_{k=1}^{N} S_{B_k}(\omega + \omega_k) + S_{B_k}(\omega - \omega_k)\right]$$
$$+ \frac{1}{4}\left[\sum_{k=1}^{N} S_{D_k}(\omega + \omega_k) + S_{D_k}(\omega - \omega_k)\right]. \qquad (6.74)$$

In the OFDM scheme, the sub-carriers form a set of functions that are orthogonal to each other, that is, the integral of the product between any two of these functions within the interval of a symbol is null. The orthogonality ensures that the ISI in the frequencies of the sub-carriers is null.

Appendix A

Fourier and Hilbert Transforms

This appendix presents some properties of the Fourier series, and the Fourier and Hilbert transforms (Hsu, 1973; Spiegel, 1976; Baskakov, 1986; Haykin, 1987; Lathi, 1989; Gradshteyn and Ryzhik, 1990; Oberhettinger, 1990; Alencar and da Rocha Jr., 2005).

Trigonometric series: $f(t) = a_0 + \sum_{n=1}^{\infty}(a_n \cos n\omega_0 t + b_n \sin n\omega_0 t)$

Cosine series: $f(t) = C_0 + \sum_{n=1}^{\infty} C_n \cos(n\omega_0 t + \theta_n)$

Complex series: $f(t) = \sum_{n=-\infty}^{\infty} F_n e^{jn\omega_0 t}$, $f(t+T) = f(t)$, $\omega_0 = \frac{2\pi}{T}$.

Conversion formulas:

$$F_0 = a_0, \qquad F_n = \frac{1}{2}(a_n - jb_n), \qquad F_{-n} = \frac{1}{2}(a_n + jb_n),$$

$$F_n = |F_n|e^{j\phi_n}, \qquad |F_n| = \frac{1}{2}\sqrt{a_n^2 + b_n^2},$$

$$a_0 = F_0, \qquad a_n = F_n + F_{-n}, \qquad b_n = j(F_n - F_{-n}),$$

$$C_0 = a_0, \qquad C_n = 2|F_n| = \sqrt{a_n^2 + b_n^2}, \qquad \theta_n = -\arctan\left(\frac{b_n}{a_n}\right).$$

Definitions of Fourier transforms and some properties:

- Definition of the Fourier transform: $F(\omega) = \int_{-\infty}^{\infty} f(t)e^{-j\omega t}dt$.

- Inverse of the Fourier transform: $f(t) = \frac{1}{2\pi} \int_{-\infty}^{\infty} F(\omega)e^{j\omega t}d\omega$.

- Magnitude and phase of the transform: $F(\omega) = |F(\omega)|e^{j\theta(\omega)}$.

- Transform of an even function $f(t)$: $F(\omega) = 2\int_{0}^{\infty} f(t)\cos\omega t dt$.

- Transform of an odd function $f(t)$: $F(\omega) = -2j\int_{0}^{\infty} f(t)\sin\omega t dt$.

- Area under function in the time domain: $F(0) = \int_{-\infty}^{\infty} f(t)dt$.

- Area under transform in the frequency domain:
  $f(0) = \frac{1}{2\pi} \int_{-\infty}^{\infty} F(\omega)d\omega$.

- Linearity of the Fourier transform: $\alpha f(t) + \beta g(t) \leftrightarrow \alpha F(\omega) + \beta G(\omega)$.

- Definition of Hilbert transform: $\hat{f}(t) = \frac{1}{\pi} \int_{-\infty}^{\infty} \frac{f(\tau)}{t-\tau}d\tau$.

Parseval's theorem:

$$\int_{-\infty}^{\infty} f(t)g(t)dt = \frac{1}{2\pi} \int_{-\infty}^{\infty} F(\omega)G^*(\omega)d\omega,$$

$$\int_{-\infty}^{\infty} |f(t)|^2 dt = \frac{1}{2\pi} \int_{-\infty}^{\infty} |F(\omega)|^2 d\omega,$$

$$\int_{-\infty}^{\infty} f(\omega)G(\omega)d\omega = \int_{-\infty}^{\infty} F(\omega)g(\omega)d\omega.$$

Fourier Transforms

$f(t)$	$F(\omega)$
$F(t)$	$2\pi f(-\omega)$
$f(t-\tau)$	$F(\omega)e^{-j\omega\tau}$
$f(t)e^{j\omega_0 t}$	$F(\omega-\omega_0)$
$f(at)$	$\frac{1}{|a|}F(\frac{\omega}{a})$
$f(-t)$	$F(-\omega)$
$f^*(t)$	$F^*(-\omega)$
$f(t)\cos\omega_0 t$	$\frac{1}{2}F(\omega-\omega_0)+\frac{1}{2}F(\omega+\omega_0)$
$f(t)\sin\omega_0 t$	$\frac{1}{2j}F(\omega-\omega_o)-\frac{1}{2j}F(\omega+\omega_0)$
$f'(t)$	$j\omega F(\omega)$
$f^{(n)}(t)$	$(j\omega)^n F(\omega)$
$\int_{-\infty}^{t} f(x)dx$	$\frac{1}{j\omega}F(\omega)+\pi F(0)\delta(\omega)$
$-jtf(t)$	$F'(\omega)$
$(-jt)^n f(t)$	$F^{(n)}(\omega)$
$f(t)*g(t)=\int_{-\infty}^{\infty} f(\tau)g(t-\tau)dx$	$F(\omega)G(\omega)$

Fourier Transforms

$f(t)$	$F(\omega)$
$\delta(t)$	1
$\delta(t-\tau)$	$e^{-j\omega\tau}$
$\delta'(t)$	$j\omega$
$\delta^{(n)}(t)$	$(j\omega)^n$
$e^{-at}u(t)$	$\frac{1}{a+j\omega}$
$e^{-a|t|}$	$\frac{2a}{a^2+\omega^2}$
e^{-at^2}	$\sqrt{\frac{\pi}{a}}e^{-\omega^2/(4a)}$
te^{-at^2}	$j\sqrt{\frac{\pi}{4a^3}}\omega e^{-\omega^2/(4a)}$
$te^{-at}u(t)$	$\frac{1}{(a+j\omega)^2}$
$\frac{t^{n-1}}{(n-1)!}e^{-at}u(t)$	$\frac{1}{(a+j\omega)^n}$
$p_T(t) = \begin{cases} 0 & \text{for } |t|>T/2 \\ A & \text{for } |t|\leq T/2 \end{cases}$	$AT\frac{\sin(\frac{\omega T}{2})}{(\frac{\omega T}{2})}$
$\frac{\sin at}{\pi t}$	$p_{2a}(\omega)$
$e^{-at}\sin bt\, u(t)$	$\frac{b}{(a+j\omega)^2+b^2}$
$e^{-at}\cos bt\, u(t)$	$\frac{a+j\omega}{(a+j\omega)^2+b^2}$

Fourier Transforms

$f(t)$	$F(\omega)$
$\frac{1}{a^2+t^2}$	$\frac{\pi}{a}e^{-a|\omega|}$
$\frac{-t}{a^2+t^2}$	$j\pi e^{-a|\omega|}[u(-\omega)-u(\omega)]$
$\frac{\cos bt}{a^2+t^2}$	$\frac{\pi}{2a}[e^{-a|\omega-b|}+e^{-a|\omega+b|}]$
$\frac{\sin bt}{a^2+t^2}$	$\frac{\pi}{2aj}[e^{-a|\omega-b|}-e^{-a|\omega+b|}]$
$\sin bt^2$	$\frac{\pi}{2b}\left[\cos\frac{\omega^2}{4b}-\sin\frac{\omega^2}{4b}\right]$
$\cos bt^2$	$\frac{\pi}{2b}\left[\cos\frac{\omega^2}{4b}+\sin\frac{\omega^2}{4b}\right]$
$\operatorname{sech} bt$	$\frac{\pi}{b}\operatorname{sech}\frac{\pi\omega}{2b}$
$\ln\left[\frac{x^2+a^2}{x^2+b^2}\right]$	$\frac{2e^{-b\omega}-2e^{-a\omega}}{\pi\omega}$
$f(t)g(t)$	$\frac{1}{2\pi}F(\omega)*G(\omega)=\frac{1}{2\pi}\int_{-\infty}^{\infty}F(\phi)G(\omega-\phi)d\phi$
$e^{j\omega_0 t}$	$2\pi\delta(\omega-\omega_0)$
$\cos\omega_0 l$	$\pi[\delta(\omega-\omega_0)+\delta(\omega+\omega_0)]$
$\sin\omega_0 t$	$-j\pi[\delta(\omega-\omega_0)-\delta(\omega+\omega_0)]$
$\sin\omega_0 tu(t)$	$\frac{\omega_0}{\omega_0^2-\omega^2}+\frac{\pi}{2j}[\delta(\omega-\omega_0)-\delta(\omega+\omega_0)]$
$\cos\omega_0 tu(t)$	$\frac{j\omega}{\omega_0^2-\omega^2}+\frac{\pi}{2}[\delta(\omega-\omega_0)+\delta(\omega+\omega_0)]$

Fourier Transforms

$f(t)$	$F(\omega)$
$u(t)$	$\pi\delta(\omega) + \frac{1}{j\omega}$
$u(t - \tau)$	$\pi\delta(\omega) + \frac{1}{j\omega}e^{-j\omega\tau}$
$u(t) - u(-t)$	$\frac{2}{j\omega}$
$tu(t)$	$j\pi\delta'(\omega) - \frac{1}{\omega^2}$
1	$2\pi\delta(\omega)$
t	$2\pi j\delta'(\omega)$
t^n	$2\pi j^n \delta^{(n)}(\omega)$
$|t|$	$\frac{-2}{\omega^2}$
$\frac{1}{t}$	$\pi j - 2\pi ju(\omega)$
$\frac{1}{t^n}$	$\frac{(-j\omega)^{n-1}}{(n-1)!}[\pi j - 2\pi ju(\omega)]$
$\frac{1}{e^{2t}-1}$	$\frac{-j\pi}{2}\coth\frac{\pi\omega}{2} + \frac{j}{\omega}$
$f_P(t) = \frac{1}{2}[f(t) + f(-t)]$	$\text{Re}(\omega)$
$f_I(t) = \frac{1}{2}[f(t) - f(-t)]$	$j\text{Im}(\omega)$
$f(t) = f_P(t) + f_I(t)$	$F(\omega) = \text{Re}(\omega) + j\text{Im}\omega$

Fourier Transforms

$f(t)$	$F(\omega)$
$\sin\left(\frac{t^2}{4a} + \frac{1}{4}\pi\right)$	$2\sqrt{\pi a}\,\sin(a\omega^2)$
$\cos\left(\frac{t^2}{4a} - \frac{1}{4}\pi\right)$	$2\sqrt{\pi a}\,\cos(a\omega^2)$
$\dfrac{\sin(\alpha)}{\cosh(t)+\cos(\alpha)}$	$\dfrac{\pi\,\sin,(\alpha\omega)}{\sin,(\pi\omega)} \quad -\pi < \alpha < \pi$
$\dfrac{\cos(\frac{1}{2}\alpha)\,\cosh(\frac{1}{2}t)}{\cosh(t)+\cos(\alpha)}$	$\dfrac{\pi\,\cosh(\alpha\omega)}{\cosh(\pi\omega)} \quad -\pi < \alpha < \pi$
$\dfrac{\Gamma(1-s)\sin\left(\frac{1}{2}s\pi\right)}{|t|^{1-s}}$	$\pi|\omega|^{-s} \quad 0 < \mathrm{Re}\,(s) < 1$
$\dfrac{1}{|t|}$	$\dfrac{\sqrt{2\pi}}{|\omega|}$
$\dfrac{\sqrt{\sqrt{\alpha^2+t^2}+\alpha}}{\sqrt{\alpha^2+t^2}}$	$\sqrt{\dfrac{2\pi}{|\omega|}}\,e^{-\alpha|\omega|}$
$\delta_T(t) = \sum_{n=-\infty}^{\infty}\delta(t-nT)$	$\omega_0\delta_{\omega_0}(\omega) = \omega_0\sum_{n=-\infty}^{\infty}\delta(\omega-n\omega_0)$

This appendix also presents some useful Hilbert transforms. Assume, for some of the properties that the signal is bandpass (Baskakov, 1986; Haykin, 1987; Poularikas, 1999)

Hilbert Transforms

$m(t)$	$\hat{m}(t)$
α	0
$m(t)$	$\hat{m}(t) = \frac{1}{\pi t} * m(t)$
$\alpha m(t) + \beta n(t)$	$\alpha \hat{m}(t) + \beta \hat{n}(t)$
$m(t) \cos \omega t$	$m(t) \sin \omega t$
$m(t) \sin \omega t$	$-m(t) \cos \omega t$
$\cos \omega t$	$\sin \omega t$
$\cos^2 \omega t$	$\frac{1}{2} \cos 2\omega t$
$\sin \omega t$	$-\cos \omega t$
$\sin^2 \omega t$	$-\frac{1}{2} \sin 2\omega t$
$\frac{\sin \alpha t}{\alpha t}$	$\frac{1 - \cos \alpha t}{\alpha t}$
$\delta(t)$	$\frac{1}{\pi t}$
$\frac{1}{t}$	$-\pi \delta(t)$

Hilbert Transforms

$m(t)$	$\hat{m}(t)$
e^{-jt}	je^{-jt}
$e^{j\omega t}$	$-j\mathrm{sgn}(\omega)e^{j\omega t}$
e^{-t^2}	$\frac{2}{\sqrt{\pi}}F(t)$
$e^{-\alpha|t|}$	$\frac{1}{\pi}\int_0^\infty \frac{2\alpha}{\alpha^2-\omega^2}\sin(\omega t)d\omega$
$\mathrm{sgn}(t)e^{-\alpha|t|}$	$-\frac{1}{\pi}\int_0^\infty \frac{2\alpha}{\alpha^2-\omega^2}\cos(\omega t)d\omega$
$e^{-\alpha t}u(t)$	$\frac{1}{\pi}\int_0^\infty \frac{\alpha\sin(\omega t)-\omega\cos(\omega t)}{\alpha^2-\omega^2}d\omega$
$\frac{\alpha}{\alpha^2+t^2}$	$\frac{t}{\alpha^2+t^2}$
$\frac{t}{\alpha^2+t^2}$	$-\frac{\alpha}{\alpha^2+t^2}$
$u(t+T)-u(t-T)$	$\frac{1}{\pi}\ln\left|\frac{t+T}{t-T}\right|$

The Dawson function, named after H. G. Dawson, is defined as

$$F(x) = e^{-x^2}\int_0^x e^{t^2}\,dt.$$

Appendix B

Biography of the Author

Marcelo Sampaio de Alencar was born in Serrita, Brazil, in 1957. He received his Bachelor Degree in Electrical Engineering, from the Federal University of Pernambuco (UFPE), Brazil, 1980, his Master Degree in Electrical Engineering, from the Federal University of Paraiba (UFPB), Brazil, 1988 and his Ph.D. from the University of Waterloo, Department of Electrical and Computer Engineering, Canada, 1993. Marcelo S. Alencar has 38 years of engineering experience, and 28 years as an IEEE Member, currently as Senior Member. Between 1982 and 1984, he worked for the State University of Santa Catarina (UDESC). From 1984 to 2003, he worked for the Department of Electrical Engineering, Federal University of Paraiba, where he was Full Professor and supervised more than 60 graduate and several undergraduate students. From 2003 to 2017, he was Chair Professor at the Department of Electrical Engineering, Federal University of Campina Grande, Brazil. He also spent some time working for MCI-Embratel and University of Toronto, as Visiting Professor. Currently he is Visiting Chair Professor at the Department of Electrical Engineering, Federal University of Bahia.

He is founder and President of the Institute for Advanced Studies in Communications (Iecom). He has been awarded several scholarships and grants, including three scholarships and several research grants from the Brazilian National Council for Scientific and Technological Research (CNPq), two grants from the IEEE Foundation, a scholarship from the University of Waterloo, a scholarship from the Federal University of Paraiba, an achievement award for contributions to the Brazilian Telecommunications Society (SBrT), an award from the Medicine College of the Federal University of Campina Grande (UFCG) and an achievement award from the College of Engineering of the Federal University of Pernambuco, during its 110th year celebration. Marcelo S. Alencar is a laureate of the 2014 Attilio Giarola Medal.

He published over 450 engineering and scientific papers and twenty-two books: Scientific Style in English, and Cellular Network Planning, by River Publishers, Spectrum Sensing Techniques and Applications,

Information Theory, and Probability Theory, by Momentum Press, Information, Coding and Network Security (in Portuguese), by Elsevier, Digital Television Systems, by Cambridge, Communication Systems, by Springer, Principles of Communications (in Portuguese), by Editora Universitária da UFPB, Set Theory, Measure and Probability, Computer Networks Engineering (in Portuguese), Electromagnetic Waves and Antenna Theory (in Portuguese), Probability and Stochastic Processess (in Portuguese), Digital Cellular Telephony (in Portuguese), Digital Telephony (in Portuguese), Digital Television (in Portuguese) and Communication Systems (in Portuguese), by Editora Érica Ltda, History of Communications in Brazil, History, Technology and Legislation of Communications, Connected Sex (in Portuguese), Scientific Diffusion (in Portuguese), Soul Hicups (in Portuguese), Epgraf Gráfica e Editora. He also wrote several chapters for ten books. His biography is included in the following publications: Who's Who in the World and Who's Who in Science and Engineering, by Marquis Who's Who, New Providence, USA.

He was technical program chair of the 36th Brazilian Symposium on Telecommunications and Signal Processing (SBrT 2018). He was general chair of the 8th National Conference on Communications, Networks and Information Security (ENCOM 2018). He was general chair of the 13th International Symposium on Wireless Personal Multimedia Communications (WPMC'10). He was Co-Chair, International Program Committee, The Second International Conference on Mobile Ubiquitous Computing, Systems, Services, and Technologies (UBICOMM 2008), He was vice-chair of the 9th IEEE International Symposium on Spread Spectrum Techniques and Applications (ISSSTA 2006) and member of the International Program Committee of the Second International Conference on Mobile Ubiquitous Computing, Systems, Services, and Technologies (UBICOMM 2008). Member of the Organizing Committee and South and Central America Liaison for the 2004 IEEE Workshop on Signal Processing Advances in Wireless Communications. He was co-chair of the *XV Simpósio Brasileiro de Telecomunicações*, in 1997, and of the 2008 IEEE Workshop on Signal Processing Advances in Wireless Communications (SPAWC 2008). He was member of the Organizing Committee of the *VI Simpósio Brasileiro de Telecomunicações*, in 1988, member of the International Advisory Committee do International Symposium on Personal, Indoor and Mobile Radio Communications (PIMRC'95), of the 12th International Conference on Telecommunications (ICT 2005), member of the Technical Program Committee of the Global Communications Conference (Globecom'99) and of several editions of the *Simpósio Brasileiro de Telecomunicações*.

He was TPC Chair of the International Telecommunication Symposium (ITS 2002), of the IEEE International Communications Conference (ICC'2002), of the 5th International Conference on Wireless Personal Multimidia Communications (WPMC'02), of the *Simpósio Brasileiro de Microondas e Optoeletrônica* (SBMO 2002), of the IEEE Global Communications Conference (Globecom'03), of the International Microwave and Optoelectronics Conference (IMOC 2003), of the IEEE International Communications Conference (ICC'2004), of the Wireless Communications Symposium, of the Global Communications Conference (Globecom'03), of the IEEE Wireless Communications and Networking Conference (WCNC'2004), of the International Workshop on Telecommunications (IWT 2004), of the Internatioal Wireless Communications and Computing Conference (IWCCC 2006), of the *XXXIII Congresso Brasileiro de Ensino de Engenharia* (COBENGE 2005), of the Advanced Industrial Conference on Telecommunications (AICT 2006), International Workshop on Telecommunications (IWT 2007), of the *XII Simpósio Brasileiro de Microondas e Optoeletrônica* (MOMAG 2006), of the 8th International Symposium on Syss tems and Information Security (SSI'06), of the Third Advanced International Conference on Telecommunications (AICT'07), of the The First International Conference on Mobile Ubiquitous Computing, Systems, Services, and Technologies (UBICOMM 2007), of the Fourth Advanced International Conference on Telecommunications (AICT 2008), of the International Microwave and Optoelectronics Conference (IMOC 2007), of the Wireless Communications & Networking Conference (WCNC 2008), of the Technical Committee of the International Workshop on Telecommunications (IWT 2009), of the 2009 IEEE Wireless Communications and Networking Conference (WCNC 2009), of the 16th International Conference on Telecommunications (ICT'09) of the First International Conference on Advances in Future Internet (AFIN 2009).

He was Member of the Technical Program Committee, Fourth Advanced International Conference on Telecommunications (AICT 2008), Member of the Panel Wireless Sensor Networks and Cooperative Objects, 10th International Conference on Wireless Personal Multimidia Communications (WPMC'07), Member of the Best Paper Committee, IST Mobile and Wireless Communication Summit, Member of the Technical Programm Committee, XXVI Brazilian Telecommunications Symposium (SBrT 2008), Member of the Technical Program Committee, International Microwave and Optoelectronics Conference (IMOC 2007), Member of the Technical Program Committee, Wireless Communications & Networking Conference (WCNC 2008), Member of the Technical Program Committee, International Workshop on Telecommunications (IWT 2009). Member of thee Technical Program Committee, 2009 IEEE Wireless Communications and Networking

Conference (WCNC 2009), Member of the Technical Program Committee, ITU-T Kaleidoscope, Member of the Technical Program Committee, 16th International Conference on Telecommunications (ICT'09).

He was also Member of the Technical Program Committee, XXVII Brazilian Telecommunications Symposium (SBrT 2009), Member of the Technical Program Committee. The First International Conference on Advances in Future Internet (AFIN 2009), Member of the Technical Program Committee 17th International Conference on Telecommunications (ICT 2010), Member of the Technical Program Committee (TPC), Wireless Communications & Networking Conference (WCNC 2010), Member of the Technical Program Committee (TPC), International Workshop on Telecommunications (IWT 2011), Member of the Technical Program Committee (TPC), International Telecommunication Symposium (ITS 2010), Member of the Technical Program Committee (TPC), The 9th International Information and Telecommunication Technologies Symposium (I2TS 2011), Member of the Program Committee, Brazilian Telecommunications Symposium (SBrT 2011), Member of the Program Committee, 10th I2TS – International Information and Telecommunication Technologies Conference, Member of the Technical Program Committee, International Symposium on Personal, Indoor and Mobile Radio Communications (PIMRC'11), Member of the Technical Program Committee, International CC onference on Advances in Computing, Communications and Informatics (ICACCI-2012), Member of the Program Committee, International Conference on Information Communication Technology (ICT-EurAsia 2013), Member of the Technical Program Committee, 2013 IEEE Symposium on Computers & Informatics (ISCI 2013), Member of the Technical Program Committee, 2013 IEEE Conference on Wireless Sensors (ICWiSe20133), Organizer of the Special Session "Education on Telecommunications", Brazilian Telecommunications Symposium (SBrT2013), Member of the Technical Program Committee, XVI Brazilian Microwave and Optoelectronics Symposium (MOMAG 2014). Member of the Technical Committee, International Conference on Advances in Computing, Communications and Informatics (ICACCI-2014), Organizer of the Special Session "Signal Processing for Cognitive Radio Networks", 22nd European Signal Processing Conference EUSIPCO 2014, Member of the Organizing Committee, 9th European Conference on Antennas & Propagation (EuCAP 2015), Member of the Technical Program Committee, 21st International Conference on Telecommunications (ICT 2014), Member of the Technical Program Committee, 2014 International Workshop on the Design and Performance of Networks on Chip (DPNoC 2014). Member of the Technical Program Committee, International Symposium on Signal Processing and Intelligent Recognition Systems (SIRS-2014). Organised by

Indian Institute of Information Technology and Management-Kerala (IIITM-K), Member of the Technical Program Committee, International Workshop on Mobile Applications (MobiApps 2014).

Marcelo S. Alencar has contributed in different capacities to the following scientific journals: Editor of the Journal of the Brazilian Telecommunication Society; Member of the International Editorial Board of the Journal of Communications Software and Systems (JCOMSS), published by the Croatian Communication and Information Society (CCIS); Member of the Editorial Board of the Journal of Networkss (JNW), published by Academy Publisher; Editor-in-Chief of the Journal of Communication and Information Systems (JCIS), special joint edition of the IEEE Communications Society (ComSoc) and SBrT. He is member of the SBrT-Brasport Editorial Board. He has been involved as a volunteer with several IEEE and SBrT activities, including being a member of the Advisory or Technical Program Committee in see veral events. He served as member of the IEEE Communications Society Sister Society Board and as liaison to Latin America Societies. He also served on the Board of Directors of IEEE's Sister Society SBrT. He is a Registered Professional Engineer. He is a columnist of the traditional Brazilian newspaper Jornal do Commercio, since April, 2000, and was Vice-President External Relations of SBrT. He is member of the IEEE, IEICE, in Japan, SBrT, SBMO, SBPC, ABJC and SBEB, in Brazil.

Appendix C

Glossary

This glossary presents the common acronyms and terminology used in modulation theory and communications (Furiati, 1998; Alencar, 1999; Design, 2001; MobileWord, 2001; Skycell, 2001; TIAB2B.com, 2001; Alencar and da Rocha Jr., 2005; Alencar, 2009a; Alencar, 2011; Alencar et al., 2018).

ADPCM – *Adaptive Differential Pulse Code Modulation*. ADPCM is a speech coding method that achieves bit rate reduction using adaptive prediction and adaptive quantization.

ADSL – *Asymmetric Digital Subscriber Line*. A modem that permits the transmission of digital information in a twisted pair.

A/D – *Analog-to-Digital Converter*. A device that converts from the analog domain to the discrete domain.

AF – *Audio Frequency*. The range of frequencies, approximately 20 Hz to 20 kHz, that when transmitted as acoustic waves, can be heard by the normal human ear, when transmitted as acoustic waves.

AGC – *Automatic Gain Control*. A function of the receiver that produces a constant power output, even if the input is varying, for instance, because of fading.

AM – *Amplitude Modulation*. In a modulation system, in which the modulating signal controls the carrier amplitude.

AMPS – *Advanced Mobile Phone Service*. The first-generation analog cellular phone system that originated in the United States.

AM-SC – *Amplitude Modulation Suppressed-Carrier*. A modulation scheme in which the unmodulated carrier is eliminated to save energy.

AM-VSB – *AM-Vestigial Sideband Modulation*. Classical method used to modulate carriers in traditional analog video.

ANSI – *American National Standards Institute.* American National Standards Institute, a US-based organization which develops standards and defines interfaces for telecommunications.

ARQ – *Automatic ReQuest for retransmission.* A type of communication link, in which the receiver asks the transmitter to re-send a block of data when errors are detected.

ASCII – *American Standard Code for Information Interchange.* ASCII data is a standard seven-bit code with one parity bit. ASCII data can be interchanged between almost every type of computer.

ASK – *Amplitude Shift Keying.* A modulation scheme in which the modulating signal changes the carrier amplitude.

ATSC – *Advanced Television System Committee.* The American forum to discuss digital television.

AWGN – *Additive White Gaussian Noise.* The usual type of noise that affects signal transmission in a channel, usually produced by thermal effect.

BCH – *Bose-Chaudhuri-Hocquenghem.* An error correction code used in the AMPS standard.

BER – *Bit Error Rate.* A measure of the error introduced in the signal, computed at the receiver end.

BPSK – *Binary Phase Shift Keying.* A binary phase modulation scheme.

BSS – *Broadcasting Satellite Service.*

BTA – *Broadcasting Technology Association (Japanese industrial organization).*

BW – *Bandwidth.* The difference between the limiting frequencies within which performance of a device, regarding some characteristic, falls within specified limits.

CIF – *Common Interchange Format.* A standardized format for the picture resolution, frame rate, color space, and color sub-sampling of digital video sequences, used in video systems.

C-QUAM – *Compatible QUAM.* An AM-stereo radio system developed by Motorola.

CATV – *Cable TV*. A television distribution method in which signals from distant stations are received, amplified, and then transmitted by (coaxial or fiber) cable or microwave links to users.

Cable modem – Modem used with optical fiber of coaxial cable.

CCI – *Cochannel Interference*. Superposition of adjacent signals in the frequency spectrum.

CCITT – *Comité Consultatif International de Telegraphique et Telephonique*. The Consultative Committee on International Telephone and Telegraph is an international organization which develops standards and defines interfaces for telecommunications (ITU-T)

CDMA – *Code Division Multiple Access*. A multiple access system that uses spread spectrum.

CELP – *Code-Excited Linear Prediction*. An analog-to-digital voice coding scheme.

CENELEC – *Comité Européen de Normalisation Électrotechnique*.

C/I – *Carrier-to-Interference Ratio*. The ratio of the desired unmodulated signal power to the interfering signal power.

CNR – *Carrier-to-Noise Ratio*. The ratio of the level of the carrier to that of the noise in the intermediate frequency (IF) band before any nonlinear process, such as amplitude limitation and detection, takes place.

Codec – *Coder/decoder*. Device that converts analogue signals to digital signals and vice versa.

COFDM – *Coded Orthogonal Frequency Division Multiplex*. A frequency multiplexing system that uses coded signals as inputs.

CRC – *Cyclic Redundancy Check*. A method of detecting errors in the serial transmission of data. A CRC for a block of data is calculated before it is sent, and is then sent along with the data. A new CRC is calculated on the received data. If the new CRC does not match the one that has been sent along with the data then an error has occurred.

D/A – *Digital-to-Analog converter*. A device to convert from the discrete to the analog domain.

dB – Decibel. Abbreviation for decibel(s). One-tenth of the common logarithm of the ratio of relative powers, equal to 0.1 B (bel).

dBi – Unit used to express the gain related to the isotropic antenna.

dBm – Decibel relates to 1 milliwatt.

DCT – *Discrete Cosine Transform.*

DEMUX – *Demultiplexer.* The separation of two or more channels previously multiplexed. *i.e.*, the reverse of multiplexing.

DFE – *Decision Feedback Equalizer.*

DPCM – *Differential Pulse Code Modulation.* Form of pulse code modulation for which efficiency is enhanced by transmitting the difference between the current signal strength and the previous pulse signal strength rather than the absolute values.

DQPSK – *Differential Quadrature Phase Shift Keying.*

DS-SS – *Direct Sequence Spread Spectrum.* A type of spread spectrum technique that uses a pseudo-random sequence.

DSL – *Digital Subscriber Line.* A dedicated link using leased line or wireless for subscriber connection.

DTMF – *Dual-Tone Multi-Frequency.* The signalling scheme used in "touch tone" telephones. Each depressed key generates two audio tones in this scheme.

DTV – *Digital Television.* An acronym for any digital television standard.

DVB – *Digital Video Broadcasting.* An acronym for the European digital television standard.

DVD – *Digital Video Disk.* An acronym for the digital video standard.

Eb/No – *Energy-per-bit to noise density ratio.* A ratio that is used to compare digital modulation schemes.

EDTV – *Enhanced Definition Television.* The quality of EDTV is between SDTV and HDTV.

ETSI – *European Telecommunications Standards Institute.* The board that has the authority to produce international communications standards in Europe.

FCC – *Federal Communications Commission.* The US Government board that has the authority to regulate all non-Federal Government interstate telecommunications (including radio and television broadcasting) as well as all international communications that originate or terminate in the United States.

FDMA – *Frequency Division Multiple Access.*

FEC – *Forward Error Correction.* System of error control for data transmission for which the receiving device has the capability to detect and correct any character or code block that contains fewer than a predetermined number of symbols in error.

FFT – *Fast Fourier Transform.* An efficient algorithm devised by John Tukey and J. W. Cooley to compute the Fourier transform of a discrete signal.

FH-SS – *Frequency-Hopping Spread Spectrum.* A type of spread spectrum technique in which the frequency of the carrier changes pseudo-randomly.

FM – *Frequency Modulation.* A modulation scheme in which the carrier frequency is a linear function of the modulating signal.

FSK – *Frequency Shift Keying.* A digital modulation technique in which the modulating signal changes the carrier frequency.

FTP – *File Transfer Protocol.* File transfer protocol is the TCP/IP standard for remote file transfer.

FVC – *Forward Voice Channel.* The radio channel used for communication of voice and user data from the base station to a cellular device.

GEO – *Geo-stationary Earth Orbit.* Communications system with satellites in geosynchronous orbits 40,000 km miles above the Earth, on the Equator plane.

GHz – Gigahertz. A unit of frequency denoting 10^9 Hz.

GMSK – *Gaussian Minimum Shift Keying.* A modulation scheme that uses a Gaussian pulse.

GSM – *Global System for Mobile Communication.* GSM originally stood for Groupe Speciale Mobile, but has been renamed to Global System for Mobile Communications, an international digital cellular standard.

HDSL – *High Data rate Digital Subscriber Line.*

HDTV – *High Definition Television.* Television that has approximately twice the horizontal and twice the vertical emitted resolution specified by the NTSC standard.

HEO – *Highly Elliptical Orbit.* This class of satellites covers orbits which have large eccentricities (are highly elliptical).

HF – *High Frequency.* From 3 to 30 MHz.

HFC – Hybrid network, including fiber and coaxial cable.

Hz – *Hertz.* Frequency unit equivalent to 1 cycle per second.

IEEE – *Institute of Electrical and Electronic Engineers.*

IEC – *International Electrotechnical Commission*

IF – *Intermediate Frequency.* A frequency to which a carrier frequency is shifted as an intermediate phase in transmission or reception.

IFFT – *Inverse Fast Fourier Transform.*

ISB – *Independent Sideband.* Double-sideband transmission in which the information carried by each sideband is different.

ISDB – *Integrated Services Digital Broadcasting.*

ISDN – *Integrated Services Digital Network.* A digital phone service which provides two data channels, each with its own phone number

ISDTV – *International System for Digital Television.* HDTV standard developed in Brazil and based on the ISDB standard.

ISI – *Intersymbol Interference.* Digital communication system impairment in which adjacent symbols in a sequence are distorted, creating dispersion that interferes in the time domain with neighboring symbols.

ISO – *International Organization for Standardization*

ITU – *International Telecommunications Union.* A United Nations (UN) agency, headquartered in Geneva, Switzerland, specialized in producing recommendations for communication systems.

ITU-R – *International Telecommunications Union, Radiocommunication Sector.*

ITU-T – *International Telecommunications Union, Telecommunication Standardisation Sector.*

JPEG – *Joint Photographic Experts Group.* A method of lossy compression for digital images.

LAN – *Local Area Network.* A computer network limited to the immediate area, usually the same building or floor of a building.

LEO – *Low Earth Orbit.* Mobile communications satellite between 700 and 2,000 kilometers above the Earth.

LF – *Low Frequency.* Frequency band between 30 kHz and 300 kHz.

LMDS – *Local Multipoint Distribution System.*

MCM – *Multicarrier Modulation.* A technique of transmitting data by dividing it into several interleaved bit streams and using these to modulate several carriers.

MEO – *Medium Earth Orbit.* MEO satellites orbit about 10,000 km above the earth.

ML – *Maximum likelihood.*

Modem – *Modulator/demodulator.* A device that can encode digital signals from a computer into analog signals that can be transmitted over analog phone lines, and vice versa.

MPEG – *Motion Picture Experts Group.*

MPSK – *M-ary Phase Shift Keying.* A modulation scheme in which the carrier phase is a function of the information signal.

MSE – *Mean Square Error.*

MSK – *Minimum Shift Keying* A modulation scheme in which the carrier frequency is a function of the information signal, and the pulse format has the sine shape.

MUX – *Multiplexer*. A device that combines multiple inputs into an aggregate signal to be transported via a single transmission channel.

NBFM – *Narrow-Band Frequency Modulation*. A scheme with a low modulation index.

NEXT – *Near-end crosstalk*. Impairment typically associated with twisted-pair transmission, in which a local transmitter interferes with a local receiver.

NF – *Noise Figure*. A measure of how much noise is produced by a device.

NRZ – *Not-Return-to-Zero*. Data encoding format in which each bit is represented for the entire duration of the bit period as a logic high or a logic low.

OFDM – *Orthogonal Frequency Division Multiplexing*. Multi-carrier signaling technique designed to maximize throughput in channels with potentially poor frequency response.

OQPSK – *Offset Quadrature Phase Shift Keying*. QPSK system in which the two bits that compose a QPSK symbol are offset in time by a half-bit period for nonlinear amplification.

PABX – *Private Automatic Branch eXchange*.

PAL – *Phase Alternated Line*. A television signal standard (625 lines, 50 Hz, 220 V primary power) used in the United Kingdom, much of the rest of western Europe, several South American countries, some Middle East and Asian countries, several African countries, Australia, New Zealand, and other Pacific island countries.

PAL-M – A modified version of the phase-alternation-by-line (PAL) television signal standard (525 lines, 50 Hz, 220 V primary power), used in Brazil.

PAM – *Pulse Amplitude Modulation*. Amplitude modulation of a carrier which uses pulses of varying amplitude to transmit information from source to destination

PCM – *Pulse Code Modulation*. A common way of converting an analog signal to a digital signal. This is done by sampling the signal and coding the sample.

PCS – *Personal Communications Service*.

PLL – *Phase-Locked Loop.* A tracking system that is used to recover a modulated carrier.

PM – *Phase Modulation.* A modulation system in which the carrier phase is a linear function of the modulating signal.

PMD – *Principle of Majority Decision.* The most frequent symbol is chosen in a detection system.

PN – *Pseudo-random Noise.* An interfering signal produced by a shift register, that emulates noise.

PSK – *Phase Shift Keying.* A modulation system in which the carrier phase is a function of the discrete modulating signal.

PSTN – *Public Switched Telephone Network.* The traditional voice network infrastructure, including both local service and long distance service.

QAM – *Quadrature Amplitude Modulation.*

QCIF – *Quarter Common Interchange Format.* A common interchange format, in which the height and width of the frame are halved.

QoS – *Quality of Service.* A parameter used to describe the attributes of a variety of network functions.

QPSK – *Quadrature Phase Shift Keying.*

QUAM – *Quadrature Amplitude Modulation.*

RACE – *Research and Development for Advanced Communications.* A European Community endeavor aimed at creating advanced communications networks.

RF – *Radio Frequency.*

RGB – *Red-Blue-Green.* Related to the use of three separate signals to carry the red, green, and blue components, respectively, of a color video image.

RLE – *Run-Length Encoding.*

RS – *Reed-Solomon (code).*

Rx – *Receiver.*

RZ – *Return-to-Zero*. Data encoding format in which each bit is represented for only a portion of the bit period as a logic high or a logic low, and what remains of the duration of the bit returns to logic zero.

SAN – *Small Area Network*. A network, generally limited to tens of meters, which uses specialized communications methods and is applied in such areas as process control and other specific real time computer applications.

S-CDMA – *Synchronous Code Division Multiple Access*.

SDH – *Synchronous Digital Hierarchy*. The ITU standard for synchronous transmission.

SDMA – *Space division Multiple Access*. A multiple access scheme that relies on intelligent antennas to separate the transmitted signals.

SDTV – *Standard Definition Television*. A digital television standard whose resolution is equivalent to the analog television.

SHF – *Super High Frequency*. Frequency band from 3 to 30 GHz.

SNR – *Signal-to-noise ratio*. A metric used to compare modulation schemes.

SONET – *Synchronous Optical Network*. The US standard for digital optical networks.

SQNR – *Signal-to-Quantization Noise Ratio*. A metric used to compare quantization schemes.

SSB – *Single Sideband*. A modulation scheme that transmits only of of the two sidebands, and saves bandwidth.

TACS – *Total Access Communication System*. A British analog mobile telephone standard based on the US AMPS system.

TCM – *Trellis-coded modulation*.

TCP/IP – *Transmission Control Protocol/Internet Protocol*. This is the suite of protocols that defines the Internet. Originally designed for the UNIX operating system, TCP/IP software is now available for every major kind of computer operating system.

TDM – *Time Division Multiplexing*

TDMA – *Time Division Multiple Access.*

TIA – *Telecommunications Industry Association.* A membership organization concerned with various standards aspects of the telecommunications industry.

TWTA – *Travelling Wave Tube Amplifier.* An amplifier technology designed for generation of very high-power microwave signals, such as those used in satellite communication applications.

Tx – Transmitter.

UHF – *Ultra-High Frequency.* Frequency band from 300 MHz to 3 GHz.

ULF – *Ultra-Low Frequency.* Frequency band from 300 Hz to 3,000 Hz.

VCO – *Voltage-Controlled Oscillator.* A device in which the output frequency is a function of the input voltage.

VDSL – *Very high-speed Digital Subscriber Line.* A twisted-pair technology that devised to attain higher data transmission rates than ADSL in exchange for shorter guaranteed distances.

VHF – *Very-High Frequency.* Frequency band from 30 Hz to 300 Hz.

VLF – *Very-Low Frequency.* Frequency band from 3 Hz to 30 Hz.

VSWR – *Voltage Standing Wave Ratio.* In a transmission line, the ratio of maximum to minimum voltage in a standing wave pattern.

WAN – *Wide-Area Network.* A network which covers a larger geographical area than a LAN and where telecommunications links are implemented.

WAP – *Wireless Application Protocol.* A global protocol for wireless systems to permit interaction between data services.

WBFM – *Wide-Band Frequency Modulation.* A modulation that uses a high deviation index.

WDM – *Wavelength Division Multiplexing.* A multiplexing scheme for the light frequency range.

WLL – *Wireless Local Loop.* A wireless system meant to bypass a local landline telephone system.

X.25 – A standard for packet transmission. A set of ITU protocols for long distance networks.

X.400 – A standard for packet transmission. A suite of ITU recommendations that define standards for data communication networks for message handling systems.

References

Alencar, M. S., "Measurement of the Probability Density Function of Communication Signals." In *Proceedings of the IEEE Instrumentation and Measurement Technology Conference – IMTC'89*, pp. 513–516, IEEE, Washington, DC, USA (1989).

Alencar, M. S., "A Frequency Domain Approach to the Optimization of Scalar Quantizers." In *Proceedings of the IEEE International Symposium on Information Theory*, p. 440, IEEE, San Antonio, TX, USA (1993).

Alencar, M. S., *Principles of Communications (Portuguese)*. University Publishers, UFPB, João Pessoa, Brazil (1999).

Alencar, M. S., *Sistemas de Comunicações*. Editora Érica Ltda., ISBN 85-7194-838-0, São Paulo, Brazil (2001).

Alencar, M. S., *Digital Television Systems*. Cambridge University Press, ISBN-10: 0521896029, ISBN-13: 9780521896023, Cambridge, UK (2009a).

Alencar, M. S., *Probabilidade e Processos Estocásticos*. Editora Érica Ltda., ISBN 978-85-365-0216-8, São Paulo, Brazil (2009b).

Alencar, M. S., *Telefonia Digital, Quinta Edição*. Editora Érica Ltda., ISBN 978-85-365-0312-7, São Paulo, Brazil (2011).

Alencar, M. S., *Televisão Digital, Segunda Edição*. Editora Érica Ltda., ISBN 978-85-365-0148-2, São Paulo, Brazil (2012).

Alencar, M. S., *Information Theory*. Momentum Press, LLC, ISBN-13: 978-1-60650-528-1 (print), ISBN-13: 978-1-60650-529-8 (e-book), New York, NY, USA (2015).

Alencar, M. S., and Alencar, R. T., *Probability Theory*. Momentum Press, LLC, ISBN-13: 978-1-60650-747-6 (print), New York, NY, USA (2016).

Alencar, M. S., Carvalho, F. B. S., and Lopes, W. T. A., *Spectrum Sensing Techniques and Applications*. Momentum Press, LLC, ISBN-13: 978-1-60650-979-1 (print), New York, NY, USA (2018).

Alencar, M. S., and da Rocha Jr., V. C., *Communication Systems*. Springer, ISBN 0-387-25481-1, Boston, MA, USA (2005).

Alencar, M. S., and Neto, B. G. A., "Estimation of the Probability Density Function by Spectral Analysis: A Comparative Study." In *Proceedings of the Treizième Colloque sur le Traitement du Signal et des Images – GRETSI*, pp. 377–380, GRETSI, Juan-Les-Pins, France (1991).

Alencar, M. S., and Queiroz, W. J. L., *Ondas Eletromagnéticas e Teoria de Antenas*. Editora érica Ltda., ISBN 978-85-365-0270-0, São Paulo, Brazil (2010).

Alencar, M. S., and Scharcanski, J., "Near Optimum Filtering of Quantized Signals." In *Proceedings of the IEEE International Symposium on Information Theory*, p. 117, Whistler, Canada (1995).

Armstrong, E. H., "A Method of Reducing Disturbances in Radio Signaling by a System of Frequency Modulation." In *Proceedings of the IRE*, vol. 24, pp. 689–674, IEEE (1936).

Baskakov, S. I., *Signals and Circuits*. Mir Publishers, Moscow, USSR (1986).

Blachman, N. M., and McAlpine, G. A., "The Spectrum of a High-Index FM Waveform: Woodward's Theorem Revisited." In *Proceedings of the IEEE Transactions on Communications Technology*, vol. 17, pp. 201–208, IEEE (1969).

Blake, I. F., *An Introduction to Applied Probability*. Robert E. Krieger Publishing Co., Malabar, FL, USA (1987).

Boutros, J., and Viterbo, E., "Signal Space Diversity: A Power- and Bandwidth-Efficient Diversity Technique for the Rayleigh Fading Channel." In *Proceedings of the IEEE Transactions on Information Theory*, vol. 44, pp. 1453–1467, IEEE (1998).

Boyer, C., *History of Mathematics (Portuguese)*. Edgard Blucher Publishers Ltd., São Paulo, Brazil (1974).

Carlson, B. A., *Communication Systems*. McGraw-Hill, Tokyo, Japan (1975).

Carson, J. R., "Notes on the Theory of Modulation." In *Proceedings of the IRE*, vol. 10, pp. 57–64, IEEE (1922).

Davenport, W. B., and Root, W. L., *An Introduction to the Theory of Random Signals and Noise*. Wiley-IEEE Press, New York, NY, USA (1987).

Design, C. S., *Communication Systems Design – Reference Library: Glossary*. Available at: http://www.csdmag.com/glossary (2001).

Divsalar, D., and Simon, M. K., "The Design of Trellis Coded MPSK for Fading Channels: Performance Criteria." In *Proceedings of the IEEE Transactions on Communications*, vol. 36, pp. 1004–1012, IEEE (1988).

El-Tanany, M. S., Wu, Y., and Házy, L., "Analytical Modeling and Simulation of Phase Noise Interference in OFDM-Based Digital Television Terrestrial Broadcasting Systems." In *Proceedings of the IEEE Transactions on Broadcasting*, vol. 47, pp. 20–31, IEEE (2001).

Furiati, G., "Serviços são Reclassificados." In *Proceedings of the Revista Nacional de Telecomunicações*, vol. 227, pp. 32–35 (1998).

Gagliardi, R., *Introduction to Communication Engineering*. John Wiley & Sons, New York, NY, USA (1978).

Gagliardi, R. M., *Introduction to Communications Engineering*. Wiley, New York, NY, USA (1988).

Gersho, A., "Principles of Quantization." In *Proceedings of the IEEE Transactions on Circuits and Systems*, vol. 25, pp. 427–436, IEEE (1978).

Gradshteyn, I. S., and Ryzhik, I. M., *Table of Integrals, Series, and Products*. Academic Press, Inc., San Diego, CA, USA (1990).

Gray, R. M., "Oversampled Sigma-Delta Modulation." In *Proceedings of the IEEE Transactions on Communications*, vol. 35, pp. 481–489, IEEE (1987).

Gray, R. M., "Spectral Analysis of Quantization Noise in a Single-Loop Sigma-Delta Modulator with dc Input." In *Proceedings of the IEEE Transactions on Communications*, vol. 37, pp. 588–599, IEEE (1989).

Gray, R. M., "Quantization Noise Spectra." In *Proceedings of the IEEE Transactions on Information Theory*, vol. 36, pp. 1220–1244, IEEE (1990).

Haykin, S., *Communication Systems*. Wiley Eastern Limited, New Delhi, India (1987).

Haykin, S., *Digital Communications*. John Wiley and Sons, New York, NY, USA (1988).

Haykin, S. S., *An Introduction to Analog and Digital Communications*. John Wiley, New York, NY, USA (1989).

Hsu, H. P., *Fourier Analysis (Portuguese)*. Livros Técnicos e Cientficos Publishers Ltd., Rio de Janeiro, Brazil (1973).

James, B. R., *Probability: An Intermediate Level Course (Portuguese)*. Institute of Pure and Applied Mathematics – CNPq, Rio de Janeiro, Brazil (1981).

Jeličic', B. D., and Roy, S., "Design of Trellis Coded QAM for Flat Fading and AWGN Channels." In *Proceedings of the IEEE Transactions on Vehicular Technology*, vol. 44, pp. 192–201, IEEE (1995).

Kennedy, R. S., *Fading Dispersive Communication Channels*. Wiley Interscience, New York, NY, USA (1969).

Kerpez, K. J., "Constellations for Good Diversity Performance." In *Proceedings of the IEEE Transactions on Communications*, vol. 41, pp. 1412–1421, IEEE (1993).

Knopp, K., *Theory and Application of Infinite Series*. Dover Publications, Inc., New York, NY, USA (1990).

Koufalas, P., *State Variable Approach to Carrier Phase Recovery and Fine Automatic Gain Control on Flat Fading Channels*. Master's thesis, University of South Australia, Adelaide, SA, Australia (1996).

Ladebusch, U., and Liss, C., "Terrestrial DVB (DVB-T): A Broadcast Technology for Stationary Portable and Mobile Use." In *Proceedings of the IEEE Proceedings*, vol. 94, pp. 183–193, IEEE (2006).

Lathi, B. P., *An Introduction to Random Signals and Communication Theory*. International Textbook Company, Scranton, PA, USA (1968).

Lathi, B. P., *Modern Digital and Analog Communication Systems*. Holt, Rinehart and Winston, Inc., Philadelphia, PA, USA (1989).

Lecours, M., Chouinard, J.-Y., Delisle, G. Y., and Roy, J., "Statistical Modeling of the Received Signal Envelope in a Mobile Radio Channel." In *Proceedings of the IEEE Transactions on Vehicular Technology*, vol. 37, pp. 204–212, IEEE (1988).

Lee, P. J., "Computation of the Bit Error Rate of Coherent M-ary PSK with Gray Code Bit Mapping." In *Proceedings of the IEEE Transactions on Communications*, vol. 34, pp. 488–491, IEEE (1986).

Lee, W. C. Y., *Mobile Cellular Telecommunications Systems*. McGraw-Hill Book Company, New York, NY, USA (1989).

Leon-Garcia, A., *Probability and Random Processes for Electrical Engineering*. Addison-Wesley Publishing Co., Reading, MA, USA (1989).

Lévine, B., *Fondements Théoriques de la Radiotechnique Statistique*. Éditions de Moscou, Moscow, USSR (1973).

Lipschutz, S., *Set Theory (Portuguese)*. Ao Livro Técnico S. A., Rio de Janeiro, Brazil (1968).

Lopes, W. T. A., and Alencar, M. S., "Space-Time Coding Performance Improvement Using a Rotated Constellation." In *Proceedings of the Anais do XVIII Simpósio Brasileiro de Telecomunicações (SBrT'2000)*, Gramado, RS, Brazil (2000).

Margolis, E., and Eldar, Y. C., "Nonuniform Sampling of Periodic Bandlimited Signals." In *Proceedings of the IEEE Transactions on Signal Processing*, vol. 56, pp. 2728–2745, IEEE (2008).

McMahon, E. L., "An Extension of Price's Theorem." In *Proceedings of the IEEE, PGIT*, vol. 10, pp. 168–168, IEEE (1964).

MobileWord, *Mobileword's Glossary*. Available at: http://www.mobileworld. org (2001).

Oberhettinger, F., *Tables of Fourier Transforms and Fourier Transforms of Distributions*. Springer-Verlag, Berlin, Germany (1990).

ON Semiconductor, "Balanced Modulators/Demodulators." *Mc1496, mc1496b application note, ON Semiconductor* (2018).

Paez, M. D., and Glisson, T. H., "Minimum Mean-Squared-Error Quantization in Speech PCM and DPCM Systems." In *Proceedings of the IEEE Transactions on Communications*, vol. 20, pp. 225–230, IEEE (1972).

Panta, K., and Armstrong, J., "Spectral Analysis of OFDM Signals and its Improvement by Polynomial Cancellation Coding." In *Proceedings of the IEEE Transactions on Consumer Electronics*, vol. 49, pp. 939–943, IEEE (2003).

Papoulis, A., *Probability, Random Variables, and Stochastic Processes*. McGraw-Hill, Tokyo, Japan (1981).

Papoulis, A., "Random Modulation: a Review." In *Proceedings of the IEEE Transactions on Accoustics, Speech and Signal Processing*, vol. 31, pp. 96–105, IEEE (1983a).

Papoulis, A., *Signal Analysis*. McGraw-Hill, Tokyo, Japan (1983b).

Papoulis, A., *Probability, Random Variables, and Stochastic Processes*. McGraw-Hill, New York, NY, USA (1984).

Poularikas, A. D., *The Handbook of Formulas and Tables for Signal Processing – The Hilbert Transform*. Editor Alexander D. Poularikas, CRC Press LLC, Boca Raton, FL, USA (1999).

Price, R., "A Useful Theorem for Non-Linear Devices Having Gaussian Inputs." *IRE, PGIT*, vol. 4, pp. 69–72, IEEE (1958).

Proakis, J. G., *Digital Communications*. McGraw-Hill Book Company, New York, NY, USA (1990).

Rajanna, K. M., Mahesan, K. V., and Sunanda, C., "Simulation Study and Analysis of OFDM." In *Proceedings of the International Symposium on Devices MEMS, Intelligent Systems and Communication (ISDMISC)*, pp. 5–10 (2011).

Reimers, U., DVB-T: The COFDM-based System for Terrestrial Television. *Electronics Communication Engineering Journal*, vol. 9, pp. 28–32 (1997).

Schwartz, M., *Information Transmission, Modulation, and Noise*. McGraw-Hill, New York, NY, USA (1970).

Schwartz, M., Bennett, W., and Stein, S., *Communication Systems and Techniques*. McGraw-Hill, New York, NY, USA (1966).

Schwartz, M., and Shaw, L., *Signal Processing: Discrete Spectral Analysis, Detection, and Estimation*. McGraw-Hill, Tokyo, Japan (1975).

Skrzypczak, A., Siohan, P., and Javaudin, J.-P., "Power Spectral Density and Cubic Metric for the OFDM/OQAM Modulation." In *Proceedings of the 2006 IEEE International Symposium on Signal Processing and Information Technology*, pp. 846–850, Vancouver, BC, Canada, IEEE (2006).

Skycell, *Glossary of Satellite Terminology*. Available at: http://www. satel-litetelephone.com (2001).

Slimane, S. B., "An Improved PSK Scheme for Fading Channels." In *Proceedings of the IEEE Transactions on Vehicular Technology*, vol. 47, pp. 703–710 (1998).

Spiegel, M. R., *Análise de Fourier*. McGraw-Hill do Brasil, Ltda., São Paulo, Brazil (1976).

Sripad, A. B., and Snyder, D. L., "A Necessary and Sufficient Condition for Quantization Errors to be Uniform and White." In *Proceedings of the IEEE Transactions on Accoustics, Speech and Signal Processing*, vol. 25, pp. 442–448, IEEE (1977).

Tarokh, V., Naguib, A., Seshadri, N., and Calderbank, A. R., "Space-Time Codes for High Data Rate Wireless Communication: Performance Criteria in the Presence of Channel Estimation Errors, Mobility and Multiple Paths." In *Proceedings of the IEEE Transactions on Communications*, vol. 47, pp. 199–207, IEEE (1999).

Tarokh, V., Seshadri, N., and Calderbank, A. R., "Space-Time Codes for High Data Rate Wireless Communication: Performance Criterion and Code Construction." In *Proceedings of the IEEE Transactions on Information Theory*, vol. 44, pp. 744–765, IEEE (1998).

Taub, H., and Schilling, D. L., *Principles of Communication Systems*. McGraw-Hill, Tokyo, Japan (1971).

TIAB2B.com, *Everything Communications*. Available at: http://www.tiab2b. com/glossary (2001).

van Vleck, J. H., and Middleton, D., "The Spectrum of Clipped Noise." In *Proceedings of the IEEE*, vol. 54, pp. 2–19, IEEE (1966).

van Waterschoot, T., Nir, V. L., Duplicy, J., and Moonen, M., "Analytical Expressions for the Power Spectral Density of CP-OFDM and ZP-OFDM Signals." In *Proceedings of the IEEE Signal Processing Letters*, vol. 17, pp. 371–374, IEEE (2010).

Woodward, P. M., "The Spectrum of Random Frequency Modulation." Memo. 666, Telecommunications Research Establishment, Great Malvern, England (1952).

Wozencraft, J. M., and Jacobs, I. M., *Principles of Communication Engineering*. John Wiley and Sons, New York, NY, USA (1965a).

Wozencraft, J. M., and Jacobs, I. M., *Principles of Communication Engineering*. John Wiley and Sons, New York, NY, USA (1965b).

Wylie, C. R., *Advanced Engineering Mathematics*. McGraw-Hill Book Company, London (1966).

Yamada, F., Sukys, F., Jr., G. B., Akamine, C., Raunheitte, L. T. M., and Dantas, C. E., "Sistema de TV Digital, Quinta Edição." In *Revista Mackenzie de Computação e Engenharia*. Universidade Presbiteriana Mackenzie, São Paulo, Brazil (2004).

Index